Calculus III Workbook

100 Exam Problems With Full Solutions

Covering

Introduction to Vectors

Vector Functions

Multivariable Calculus

Vector Calculus

N. Rimmer

ISBN 978-0-692-94032-7

Table of Contents

Section 2: Multivariable Calculus		
a) Multivariable Functions	Domain	2.1
b) Multivariable Derivatives	Partial Derivative	2.2, 2.3, 2.4
	Chain Rule	2.5
	Implicit Differentiation	2.6
	Directional Derivative	2.7, 2.8
	Tangent Plane	2.9, 2.10
	Linearization	2.11
	Optimization, Critical Points	2.12, 2.13, 2.14
	Lagrange Multiplier Method	2.15, 2.16, 2.17
c) Multivariable Integrals	Double Integral in Cartesian	2.18, 2.19, 2.20, 2.22
	Double Integral in Polar	2.21
	Volume as a Double Integral	2.18, 2.22, 2.23
	Triple Integral in Cartesian	2.23, 2.26, 2.29
	Triple Integral in Cylindrical	2.24, 2.27, 2.30
	Triple Integral in Spherical	2.25, 2.28, 2.31
	Volume as a Triple Integral	2.23, 2.24, 2.25, 2.26, 2.29, 2.30
	Change of Variable	2.32, 2.33, 2.34
Section 3: Vector Calculus		
a) Line Integrals	Line Integral, Parametrize	3.1, 3.2, 3.10, 3.13, 3.15, 3.27, 3.33
	Fundamental Theorem of Line Integrals	3.3, 3.4, 3.5, 3.6, 3.12, 3.14
	Green's Theorem	3.7, 3.8, 3.9, 3.11, 3.15
	Curl of a Vector Field	3.16, 3.17
	Divergence of a Vector Field	3.17
b) Surface Integrals	Surface Integral, Surface Area	3.18
	Surface Integral	3.19
	Surface Integral, Flux of a Vector Field	3.20
	Stokes' Theorem	3.21, 3.23, 3.26, 3.27, 3.28, 3.30, 3.33
	Divergence Theorem	3.22, 3.24, 3.25, 3.29, 3.31, 3.32

PREFACE

This is a collection of my Calculus III midterm exam problems. I (using methods taught during lecture) wrote the solutions. There may be an easier way to solve some of the problems, as with any question, there are multiple ways to approach the problem. If you happen to find a mistake please don't hesitate to contact me (nakiarimmer@gmail.com) to point it out. This workbook is meant for any person studying Calculus III which is normally called Multivariable Calculus. This is my second workbook of this type. In 2014 I published my Calculus II Workbook, you can find it here: **https://tinyurl.com/ycgrj4yw**. It is my hope that these workbooks will aid in learning the material. The workbook together with a good set of notes and lecture videos serve as a great education package. I am working to post lecture notes and lecture videos on my website **calccoach.com**. Be on the lookout for more study aids.

Section 1: Vectors and Vector Functions

1.1) Find the equation of the sphere that has the points $(3,2,3)$ and $(1,-4,5)$ on opposite ends of a diameter.

1.1) Find the equation of the sphere that has the points $(3,2,3)$ and $(1,-4,5)$ on opposite ends of a diameter.

The line segment connecting the points is a diameter
- Its' midpoint is the center
- $\frac{1}{2}$ of its' length is the radius

Mid point : $\left(\frac{3+1}{2}, \frac{2+-4}{2}, \frac{3+5}{2}\right) = (2,-1,4)$

$\Rightarrow$ The sphere's center is $(2,-1,4)$

$radius = \frac{1}{2}\sqrt{(3-1)^2 + (2+4)^2 + (3-5)^2} = \frac{1}{2}\sqrt{4+36+4}$

$\qquad = \frac{1}{2}\sqrt{44} = \frac{1}{2}\cdot 2\sqrt{11} = \sqrt{11}$

standard form for the equation of a sphere with center (h,k,m) and radius r

$(x-h)^2 + (y-k)^2 + (z-m)^2 = r^2$

$$\boxed{(x-2)^2 + (y+1)^2 + (z-4)^2 = 11}$$

1.2) Point A has coordinates $(-2,4,3)$

and B has coordinates $(x,-4,2)$.

If the distance from A to B is 9, what is the value(s) of x ?

If there is more than one answer, circle all that apply.

(A) -4   (E) 2
(B) -2   (F) 4
(C) -1   (G) 6
(D) -6   (H) 8

1.2) Point A has coordinates $(-2, 4, 3)$

and B has coordinates $(x, -4, 2)$.

If the distance from A to B is 9, what is the value(s) of x ?

If there is more than one answer, circle all that apply.

(A) -4   (E) 2
(B) -2   (F) 4
(C) -1   (G) 6
(D) -6   (H) 8

$$|AB| = \sqrt{(x+2)^2 + (-4-4)^2 + (2-3)^2} = 9$$

$$(x+2)^2 + 64 + 1 = 81$$

$$(x+2)^2 = 16$$

$$x + 2 = 4 \qquad\qquad x + 2 = -4$$

$$\boxed{x = 2} \qquad\qquad \boxed{x = -6}$$

1.3) For the triangle with vertices located at

$A(3,3,4)$, $B(2,2,4)$, and $C(1,1,1)$,

find a vector from the vertex C to the midpoint of AB.

a) $\left\langle \frac{3}{2}, \frac{3}{2}, 1 \right\rangle$ b) $\left\langle \frac{7}{2}, \frac{7}{2}, 4 \right\rangle$ c) $\left\langle \frac{5}{2}, \frac{5}{2}, 3 \right\rangle$ d) $\left\langle \frac{9}{2}, \frac{9}{2}, 5 \right\rangle$

e) $\left\langle \frac{1}{2}, \frac{1}{2}, 1 \right\rangle$ f) $\left\langle \frac{5}{2}, \frac{5}{2}, 4 \right\rangle$ g) $\left\langle \frac{3}{2}, \frac{3}{2}, 3 \right\rangle$ h) $\left\langle \frac{7}{2}, \frac{7}{2}, 5 \right\rangle$

1.3) For the triangle with vertices located at

$A(3,3,4)$, $B(2,2,4)$, and $C(1,1,1)$,

find a vector from the vertex C to the midpoint of AB.

a) $\left\langle \frac{3}{2}, \frac{3}{2}, 1 \right\rangle$ b) $\left\langle \frac{7}{2}, \frac{7}{2}, 4 \right\rangle$ c) $\left\langle \frac{5}{2}, \frac{5}{2}, 3 \right\rangle$ d) $\left\langle \frac{9}{2}, \frac{9}{2}, 5 \right\rangle$

e) $\left\langle \frac{1}{2}, \frac{1}{2}, 1 \right\rangle$ f) $\left\langle \frac{5}{2}, \frac{5}{2}, 4 \right\rangle$ g) $\left\langle \frac{3}{2}, \frac{3}{2}, 3 \right\rangle$ h) $\left\langle \frac{7}{2}, \frac{7}{2}, 5 \right\rangle$

$$\text{Midpoint} \atop \text{of } AB = \left(\frac{3+2}{2}, \frac{3+2}{2}, \frac{4+4}{2} \right)$$

Call the point $M\left(\frac{5}{2}, \frac{5}{2}, 4 \right)$

$$\vec{CM} = \left\langle \frac{5}{2}-1, \frac{5}{2}-1, 4-1 \right\rangle$$

$$\boxed{\vec{CM} = \left\langle \frac{3}{2}, \frac{3}{2}, 3 \right\rangle}$$

1.4) Let $A(1,0,1)$, $B(3, 1, 0)$, $C(3, 2, 2)$ and $D(-1, -2, 1)$ be vertices of the parallelepiped below.

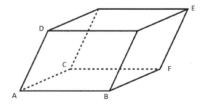

Find the volume of the parallelepiped formed by edges AB, AC, and AD.

Find the coordinates of the point E opposite to A in this parallelepiped using the fact that $\overrightarrow{AE} = \overrightarrow{AF} + \overrightarrow{AD}$.

Find the angle between $\overrightarrow{AE}$ and $\overrightarrow{AD}$.

1.4) Let $A(1,0,1)$, $B(3, 1, 0)$, $C(3, 2, 2)$ and $D(-1, -2, 1)$ be vertices of the parallelepiped below.

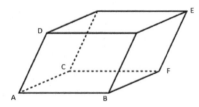

Find the volume of the parallelepiped formed by edges AB, AC, and AD.

$$\vec{AB} = \langle 3-1, 1-0, 0-1 \rangle = \langle 2,1,-1 \rangle = u$$
$$\vec{AC} = \langle 3-1, 2-0, 2-1 \rangle = \langle 2,2,1 \rangle = v$$
$$\vec{AD} = \langle -1-1, -2-0, 1-1 \rangle = \langle -2,-2,0 \rangle = w$$

$$V = |u \cdot (v \times w)| = \begin{vmatrix} 2 & 1 & -1 \\ 2 & 2 & 1 \\ -2 & -2 & 0 \end{vmatrix} = 2(2) - 1(2) - 1(0) = \boxed{2}$$

Find the coordinates of the point E opposite to A in this parallelepiped using the fact that $\vec{AE} = \vec{AF} + \vec{AD}$.

$$\vec{AF} = \vec{AB} + \vec{AC} = \langle 2+2, 2+1, 1+-1 \rangle = \langle 4,3,0 \rangle$$
$$\vec{AF} + \vec{AD} = \langle 4+-2, 3+-2, 0+0 \rangle = \langle 2,1,0 \rangle$$

Let E be (x,y,z) with $A(1,0,1)$
$$\vec{AE} = \langle x-1, y-0, z-1 \rangle \stackrel{Set}{=} \langle 2,1,0 \rangle$$

$$x-1 = 2 \Rightarrow x=3$$
$$y = 1 \qquad y=1 \qquad \boxed{E(3,1,1)}$$
$$z-1 = 0 \Rightarrow z=1$$

Find the angle between $\vec{AE}$ and $\vec{AD}$.

$$\vec{AE} = \langle 2,1,0 \rangle \quad \vec{AD} = \langle -2,-2,0 \rangle$$

$$\cos\theta = \frac{\vec{AE} \cdot \vec{AD}}{|\vec{AE}||\vec{AD}|} = \frac{-2(2)+-2(1)+0}{\sqrt{5} \cdot 2\sqrt{2}} = \frac{-6}{2\sqrt{10}}$$

$$\cos\theta = \frac{-3}{\sqrt{10}} \qquad \boxed{\theta = \cos^{-1}\left(\frac{-3}{\sqrt{10}}\right)}$$

1.5) (a) Show that the cross product of the diagonals of the parallelogram formed by vectors $\vec{u}$ and $\vec{v}$ has a magnitude twice as long as $\vec{u} \times \vec{v}$.

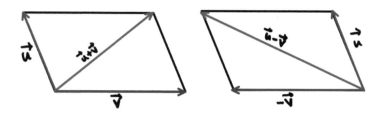

(b) A parallelogram with all sides of equal length is called a rhombus.

Using vectors, show that the diagonals of a rhombus meet in a right angle.

1.5) (a) Show that the cross product of the diagonals of the parallelogram formed by vectors $\vec{u}$ and $\vec{v}$ has a magnitude twice as long as $\vec{u} \times \vec{v}$.

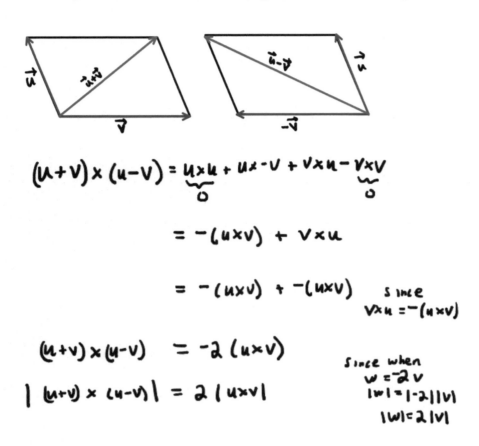

$$(u+v) \times (u-v) = \underbrace{u \times u}_{0} + u \times -v + v \times u - \underbrace{v \times v}_{0}$$

$$= -(u \times v) + v \times u$$

$$= -(u \times v) + -(u \times v) \qquad \text{since}$$
$$v \times u = -(u \times v)$$

$$(u+v) \times (u-v) = -2(u \times v)$$

$$| (u+v) \times (u-v) | = 2 | u \times v | \qquad \begin{array}{l} \text{since when} \\ w = -2v \\ |w| = |-2||v| \\ |w| = 2|v| \end{array}$$

(b) A parallelogram with all sides of equal length is called a rhombus.
Using vectors, show that the diagonals of a rhombus meet in a right angle.

To find the angle between vectors, consider the dot product between them.

$$(u+v) \cdot (u-v) = u \cdot u \underbrace{- u \cdot v + v \cdot u}_{\text{cancel}} - v \cdot v \qquad \text{since } u \cdot v = v \cdot u$$

$$(u+v) \cdot (u-v) = |u|^2 - |v|^2$$

$$(u+v) \cdot (u-v) = 0 \qquad \text{since in a rhombus } |u| = |v|$$

So u + v is orthogonal to u - v since the dot product is zero.

1.6) (a) Let $\mathbf{u} = \mathbf{i} - 2\mathbf{j}$, $\mathbf{v} = 2\mathbf{i} + 3\mathbf{j}$, and $\mathbf{w} = \mathbf{i} + \mathbf{j}$. Write $\mathbf{u} = \mathbf{u}_1 + \mathbf{u}_2$.
Where $\mathbf{u}_1$ is parallel to $\mathbf{v}$ and $\mathbf{u}_2$ is parallel to $\mathbf{w}$.

(b) If $\mathbf{a} \cdot \mathbf{b} = \sqrt{3}$ and $\mathbf{a} \times \mathbf{b} = \langle 1,1,1 \rangle$, then find the angle between $\mathbf{a}$ and $\mathbf{b}$.

1.6) (a) Let $\mathbf{u} = \mathbf{i} - 2\mathbf{j}$, $\mathbf{v} = 2\mathbf{i} + 3\mathbf{j}$, and $\mathbf{w} = \mathbf{i} + \mathbf{j}$. Write $\mathbf{u} = \mathbf{u}_1 + \mathbf{u}_2$.

Where $\mathbf{u}_1$ is parallel to $\mathbf{v}$ and $\mathbf{u}_2$ is parallel to $\mathbf{w}$.

$$u_1 = Kv \qquad u_2 = cw$$
$$u = u_1 + u_2 \Rightarrow u = kv + cw$$
$$\langle 1,-2\rangle = K\langle 2,3\rangle + c\langle 1,1\rangle$$
$$\langle 1,-2\rangle = \langle 2K+c, \ 3K+c\rangle$$
$$\Rightarrow \ 2K+c = 1$$
$$3K+c = -2$$

Solve this system of equations

$$\begin{array}{c} 2K+c = 1 \\ -(3K+c = -2) \\ \hline -K = 3 \Rightarrow K = -3 \end{array}$$

$$\begin{array}{c} 2(-3)+c = 1 \\ -6+c = 1 \\ c = 7 \end{array} \Biggr\} \quad \boxed{u = -3v + 7w}$$
$$\underset{u_1}{\underbrace{}} \quad \underset{u_2}{\underbrace{}}$$

(b) If $\mathbf{a} \cdot \mathbf{b} = \sqrt{3}$ and $\mathbf{a} \times \mathbf{b} = \langle 1,1,1\rangle$, then find the angle between $\mathbf{a}$ and $\mathbf{b}$.

$$a \cdot b = |a||b|\cos\theta \qquad\qquad |a \times b| = |a||b|\sin\theta$$
$$\underbrace{\sqrt{3}} = |a||b|\cos\theta \qquad |\langle 1,1,1\rangle| = \sqrt{1+1+1} = \sqrt{3}$$
$$\sqrt{3} = |a||b|\sin\theta$$

$$\rule{8cm}{0.4pt}$$

$$|a||b|\cos\theta = |a||b|\sin\theta \qquad\qquad \text{divide by } |a||b|$$
$$\cos\theta = \sin\theta \quad , \text{so} \quad \boxed{\theta = \frac{\pi}{4}}$$

and the angle θ b/w
vectors must be $0 \le \theta \le \pi$

1.7) (a) If $|\mathbf{v}| = 4, |\mathbf{w}| = 3$ and the angle

between $\mathbf{v}$ and $\mathbf{w}$ is $\dfrac{\pi}{3}$, find $|2\mathbf{v} - \mathbf{w}|$.

(b) Suppose $\mathbf{a}, \mathbf{b}$, and $\mathbf{c}$ are non-zero vectors. Prove :

If $\mathbf{a} \cdot \mathbf{b} = \mathbf{a} \cdot \mathbf{c}$ and $\mathbf{a} \times \mathbf{b} = \mathbf{a} \times \mathbf{c}$, then $\mathbf{b} = \mathbf{c}$.

1.7) (a) If $|\mathbf{v}| = 4, |\mathbf{w}| = 3$ and the angle

between $\mathbf{v}$ and $\mathbf{w}$ is $\dfrac{\pi}{3}$, find $|2\mathbf{v} - \mathbf{w}|$.

$$|2v-w|^2 = (2v-w)\cdot(2v-w)$$
$$= 4v\cdot v - 2v\cdot w - 2w\cdot v + w\cdot w$$
$$= 4|v|^2 - 4v\cdot w + |w|^2 \quad \text{since } v\cdot w = w\cdot v$$
$$\text{and } |v|^2 = v\cdot v$$
$$= 4\cdot 16 - 4\cdot|v||w|\cos\theta + 9 \quad \text{since } |v|=4$$
$$|w|=3$$
$$\text{and } v\cdot w = |v||w|\cos\theta$$
$$= 64 - 4\cdot 4\cdot 3\cdot\cos\tfrac{\pi}{3} + 9$$
$$= 64 - 48\cdot\tfrac{1}{2} + 9 = 64 - 24 + 9 = 49$$
$$|2v-w|^2 = 49 \quad\Rightarrow\quad \boxed{|2v-w| = 7}$$

(b) Suppose $\mathbf{a}, \mathbf{b}$, and $\mathbf{c}$ are non-zero vectors. Prove :

If $\mathbf{a}\cdot\mathbf{b} = \mathbf{a}\cdot\mathbf{c}$ and $\mathbf{a}\times\mathbf{b} = \mathbf{a}\times\mathbf{c}$, then $\mathbf{b} = \mathbf{c}$.

$$a\cdot b - a\cdot c = 0 \qquad\qquad a\times b - a\times c = 0$$
$$a\cdot(b-c) = 0 \qquad\qquad a\times(b-c) = 0$$
$$\Rightarrow a \text{ is orthogonal to } b-c \quad\Rightarrow a \text{ is parallel to } b-c$$

If a vector is parallel and orthogonal to another vector at the same time, one of them must be the zero vector.

$$a \neq 0 \quad\text{so}\quad b-c = 0 \quad\text{and thus,}\quad b = c.$$

1.8) In the figure on the below, let **u**, **v**, and **w** all be unit vectors. Find **u** · **w**.

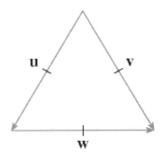

1.8) In the figure on the below, let **u**, **v**, and **w**
all be unit vectors. Find **u** · **w**.

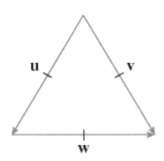

$$|u| \cdot |w| \cos\theta = u \cdot w$$

u and v are unit vectors
$$|u| = |w| = 1$$

$$\Rightarrow \cos\theta = u \cdot w$$

θ is the angle b/w
u and w

$$\theta = 60° + \alpha$$

$$\theta = 60° + 60° = 120°$$

$$\theta = \frac{2\pi}{3}$$

$$u \cdot w = \cos\left(\frac{2\pi}{3}\right) = -\frac{1}{2}$$

$$\boxed{u \cdot w = \frac{-1}{2}}$$

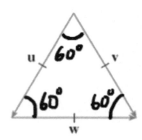

· equilateral
· equiangular

$$\alpha = 60°$$

$$\left(\begin{array}{l}\text{parallel lines} \\ \text{cut by a transversal} \\ \text{alt. int. angles are} \cong\end{array}\right)$$

1.9) Find the cosine of the acute angle
between the diagonals of a cube.

(A) $\dfrac{2}{5}$

(B) $\dfrac{2}{9}$

(C) $\dfrac{4}{9}$

(D) $\dfrac{5}{6}$

(E) $\dfrac{2}{3}$

(F) $\dfrac{1}{3}$

(G) $\dfrac{\sqrt{3}}{3}$

(H) $\dfrac{2}{\sqrt{3}}$

1.9) Find the cosine of the acute angle between the diagonals of a cube.

(A) $\dfrac{2}{5}$

(B) $\dfrac{2}{9}$

(C) $\dfrac{4}{9}$

(D) $\dfrac{5}{6}$

(E) $\dfrac{2}{3}$

(F) $\dfrac{1}{3}$

(G) $\dfrac{\sqrt{3}}{3}$

(H) $\dfrac{2}{\sqrt{3}}$

To frame the situation, make the cube a unit cube and place it in the first octant.

$A(1,0,0)$
$B(0,1,1)$ } $\vec{AB} = \langle -1, 1, 1 \rangle$

$C(1,1,0)$
$D(0,0,1)$ } $\vec{CD} = \langle -1, -1, 1 \rangle$

These vectors serve as diagonals of the cube

$|\vec{AB}| = \sqrt{1+1+1} = \sqrt{3}$

$|\vec{CD}| = \sqrt{1+1+1} = \sqrt{3}$

$\vec{AB} \cdot \vec{CD} = 1 - 1 + 1 = 1$

Let Θ be the angle b/w $\vec{AB}$ and $\vec{CD}$

$\cos\Theta = \dfrac{\vec{AB} \cdot \vec{CD}}{|\vec{AB}||\vec{CD}|} = \dfrac{1}{\sqrt{3}\cdot\sqrt{3}} = \boxed{\dfrac{1}{3}}$

1.10) Find parametric equations for the line of intersection between $x + 2y - z = 2$ and $-2x + 2y + 2z = -1$.

1.10) Find parametric equations for the line of intersection between $x+2y-z=2$ and $-2x+2y+2z=-1$.

- Check to see if the planes are parallel.

  $x+2y-z=2 \quad n_1 = \langle 1,2,-1 \rangle$

  $-2x+2y+2z=-1 \quad n_2 = \langle -2,2,2 \rangle$

  $n_1 \neq k n_2$

  $\Rightarrow$ not parallel

- Get one of the variables to cancel by using the elimination method.

  $x+2y-z=2$

  $-(-2x+2y+2z=-1)$

  $\overline{3x-3z=3} \quad div \ by \ 3 \Rightarrow x-z=1$

- Choose one of these variables to be the parameter.   let $z=t$

  Solve for x $\Rightarrow x=1+t$

  $x=1+z$

- Plug these into either plane equation to Solve for the cancelled variable.

  $x+2y-z=2$

  $1+t+2y-t=2 \Rightarrow 1+2y=2$

  $2y=1$

  $y=\frac{1}{2}$

$\boxed{\begin{array}{l} x=1+t \\ y=\frac{1}{2} \\ z=\quad t \end{array}}$

Answers may vary based on the choices made along the way.

1.11) Find the distance between the two skew lines

L_1 :

$x = 1 + t$

$y = 1 + 6t$

$z = 2t$

L_2 :

$x = 1 + 2s$

$y = 5 + 15s$

$z = -2 + 6s$

(A) 8   (C) 6   (E) 4   (G) 2

(B) 7   (D) 5   (F) 3   (H) 1

1.11) Find the distance between the two skew lines

L_1 :

$x = 1 + t$

$y = 1 + 6t$

$z = 2t$

L_2 :

$x = 1 + 2s$

$y = 5 + 15s$

$z = -2 + 6s$

(A) 8 (C) 6 (E) 4 (G) 2

(B) 7 (D) 5 (F) 3 (H) 1

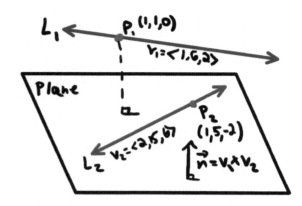

The distance between skew lines can be found by finding an equation of the plane containing one of the lines and then finding the distance between any point on the other line and that plane.

$$\vec{n} = V_1 \times V_2 = \begin{vmatrix} i & j & K \\ 1 & 6 & 2 \\ 2 & 15 & 6 \end{vmatrix} = \langle 36-30, -(6-4), 15-12 \rangle$$

$$\vec{n} = \langle 6, -2, 3 \rangle$$

- Plane: $ax + by + cz + d = 0$ use P_2

$a = 6$ $x = 1$ $6 + -10 + -6 + d = 0 \Rightarrow d = 10$

$b = -2$ $y = 5$

$c = 3$ $z = -2$ eg: $6x - 2y + 3z + 10 = 0$

- Dist. b/w this plane and P_1 :

$$d = \frac{|ax + by + cz + d|}{\sqrt{a^2 + b^2 + c^2}} = \frac{|6(1) - 2(1) + 3(0) + 10|}{\sqrt{36 + 4 + 9}} = \frac{14}{\sqrt{49}} = \boxed{2}$$

1.12) (a) Find parametric equations for the line through the point $(0,1,2)$ that is parallel to the plane $x + y + z = 2$ and perpendicular to the line

$$x \;=\; 1 \;+\; t$$
$$y \;=\; 1 \;-\; t$$
$$z \;=\; \qquad 2t$$

(b) At what point does the line in the answer to part a) intersect the plane $x = 6$?

1.12) (a) Find parametric equations for the line through the point $(0,1,2)$ that is parallel to the plane $x + y + z = 2$ and perpendicular to the line

$$x = 1 + t$$
$$y = 1 - t$$
$$z = 2t$$

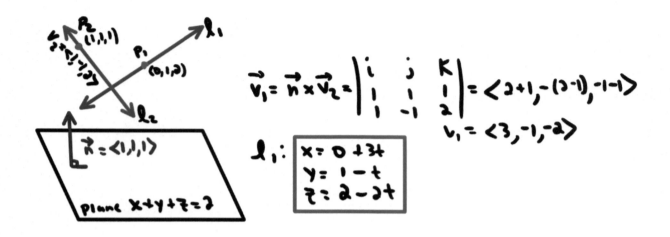

$$\vec{v_1} = \vec{n} \times \vec{v_2} = \begin{vmatrix} i & j & k \\ 1 & 1 & 1 \\ 1 & -1 & 2 \end{vmatrix} = \langle 2+1, -(2-1), -1-1 \rangle$$

$$v_1 = \langle 3, -1, -2 \rangle$$

ℓ_1:
$$x = 0 + 3t$$
$$y = 1 - t$$
$$z = 2 - 2t$$

(b) At what point does the line in the answer to part a) intersect the plane $x = 6$?

ℓ_1 intersects the plane $x = 6$: Find $t \Rightarrow 6 = 3t \Rightarrow t = 2$
plug this t in to find y & z

$$\left. \begin{array}{l} y = 1 - 2 = -1 \\ z = 2 - 4 = -2 \end{array} \right\}$$ pt: $(6, -1, -2)$

1.13) Find the distance between the parallel planes

$P_1 = x + 2y + 2z - 4 = 0$ and $P_2 = x + 2y + 2z - 10 = 0$

(A) 8   (C) 6   (E) 4   (G) 2

(B) 14   (D) 5   (F) 3   (H) 12

1.13) Find the distance between the parallel planes

$P_1 = x+2y+2z-4 = 0$ and $P_2 = x+2y+2z-10 = 0$

(A) 8   (C) 6   (E) 4   (G) 2

(B) 14   (D) 5   (F) 3   (H) 12

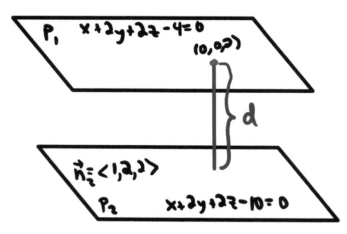

This reduces to finding the distance from a point to a plane.
Find any point on one of the planes and find its distance to the
other plane.

In P_1, set $x=y=0$ and find z. $0+0+2z=4 \Rightarrow z=2$

$$d = \frac{|ax+by+cz+d|}{\sqrt{a^2+b^2+c^2}} = \frac{|1(0)+2(0)+2(2)-10|}{\sqrt{1+4+4}} = \frac{6}{\sqrt{9}} = \boxed{2}$$

1.14) Tell whether each statement is True or False and explain your reasoning.

(*a*) Two planes parallel to the same line are parallel.

(*b*) Two lines parallel to the same plane are parallel.

1.14) Tell whether each statement is True or False
and explain your reasoning.

(a) Two planes parallel to the same line are parallel.

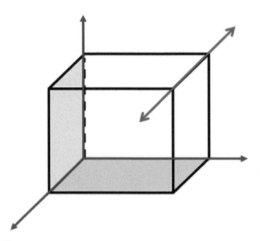

FALSE: **Take the xy-plane and the xz-plane.**
Both are parallel to the line x = t, y = 1, z = 1,
but they are orthogonal to each other

(b) Two lines parallel to the same plane are parallel.

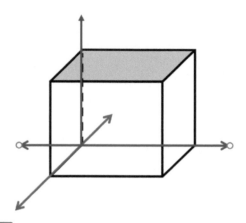

FALSE: **Take the x-axis and the y-axis.**
Both are parallel to the plane z = 1,
but they are orthogonal to each other

1.15) Find the equation of the plane that passes through $(-2,4,3)$ and contains the line

$$
\begin{aligned}
x &= -2 + 2t \\
y &= -4 + 3t \\
z &= 3 - t
\end{aligned}
$$

1.15) Find the equation of the plane that passes through $(-2, 4, 3)$ and contains the line

$$x = -2 + 2t$$
$$y = -4 + 3t$$
$$z = 3 - t$$

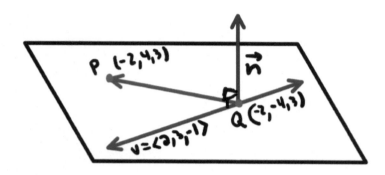

The vector from Q to P and the vector v are both in the plane. The cross product of these vectors will give the normal vector to the plane.

$$\vec{QP} = \langle -2 - 2, 4 - 4, 2 - 3 \rangle = \langle 0, 8, 0 \rangle$$

$$\vec{n} = V \times QP = \begin{vmatrix} i & j & k \\ 2 & 3 & -1 \\ 0 & 8 & 0 \end{vmatrix} = \langle 0 - 8, -(0-0), 16-0 \rangle$$

$$n = \langle 8, 0, 16 \rangle$$
reduce or scale down
$$n = \langle 1, 0, 2 \rangle$$

Plane eq.: $\left. \begin{array}{c} (-2,4,3) \\ \langle 1,0,2 \rangle \end{array} \right\}$ $\underbrace{ax + by + cz + d = 0}$

$a=1$	$x=-2$	$ax = -2$	$4 + d = 0$
$b=0$	$y=4$	$by = 0$	$d = -4$
$c=2$	$z=3$	$cz = \frac{6}{4}$	

$$\boxed{x + 2z - 4 = 0}$$

1.16) Find the plane of points that are equidistant from the points $(1,1,1)$ and $(5,-1,3)$ and determine where this plane intersects the x-axis.

a) −2   b) 0   c) 2   d) 3   e) 4   f) 6   g) 8   h) 10

1.16) Find the plane of points that are equidistant
from the points $(1,1,1)$ and $(5,-1,3)$ and determine
where this plane intersects the x-axis.

a) −2   b) 0   c) 2   d) 3   e) 4   f) 6   g) 8   h) 10

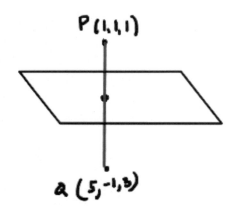

P(1,1,1)

a (5,-1,3)

The midpoint is on the plane

$$\left(\frac{1+5}{2}, \frac{1+-1}{2}, \frac{1+3}{2}\right) = (3, 0, 2)$$

The normal vector to the plane points
in the direction of the vector
b/w the points

$$\vec{PQ} = \langle 5-1, -1-1, 3-1 \rangle = \langle 4, -2, 2 \rangle$$

$$\vec{PQ} = 2\langle 2, -1, 1 \rangle$$

$$\vec{n} = \langle 2, -1, 1 \rangle$$

Now go into $ax + by + cz + d = 0$ with $a=2$ $x=3$
$b=-1$ $y=0$
$c=1$ $z=2$

$$2(3) + -1(0) + 1(2) + d = 0$$
$$6 + 0 + 2 + d = 0$$
$$d = -8$$

Find d

The plane has equation

$$2x - y + z - 8 = 0$$

A function intersects the x-axis when $y = z = 0$

$$\Rightarrow 2x - 8 = 0 \Rightarrow \boxed{x = 4}$$

1.17) Find the x coordinate of the point of the plane
$2x - y + 3z = 32$ that is closest to the point $(1,1,1)$.

a) $-1/2$ b) 0   c) $1/2$ d) 1   e) $5/2$ f) 3   g) $7/2$ h) 5

1.17) Find the x coordinate of the point of the plane

$2x - y + 3z = 32$ that is closest to the point $(1,1,1)$.

a) $-1/2$ b) 0   c) $1/2$ d) 1   e) $5/2$ f) 3   g) $7/2$ h) 5

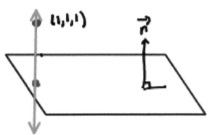

The shortest distance b/w a point and a plane
is the perpendicular distance

The line through $(1,1,1)$ and the closest pt. on the plane
has the same direction as the normal vector

$$\vec{n} = <2,-1,3>$$

Find the equation of the line from above.

$x = 1 + 2t$
$y = 1 - t$
$z = 1 + 3t$

Our answer is the intersection b/w the line and the plane.
Plug the x, y, and z from the line into the plane equation

$2(1+2t) - 1(1-t) + 3(1+3t) = 32$ Find t
$2 + 4t - 1 + t + 3 + 9t = 32$
$14t = 28 \quad t = 2$

The point of intersection is $(5, -1, 7)$
$x = 1 + 2(2) \quad y = 1 - 2 \quad z = 1 + 3(2)$

The x coordinate is $\boxed{5}$

1.18) Find parametric equations for the tangent line to the curve $\mathbf{r}(t) = \langle 2\cos(t), 2\sin(t), 4\cos(2t) \rangle$ at the point $(\sqrt{3}, 1, 2)$.

Find the point where this tangent line intersects the yz – plane.

What is the z coordinate of this point?

(A) 0   (C) 2   (E) 4   (G) −4

(B) −10   (D) 8   (F) −8   (H) $\sqrt{3}$

1.18) Find parametric equations for the tangent line to the curve

$$\mathbf{r}(t) = \langle 2\cos(t), 2\sin(t), 4\cos(2t) \rangle \text{ at the point } \left(\sqrt{3}, 1, 2\right).$$

Find the point where this tangent line intersects the yz – plane.
What is the z coordinate of this point?

(A) 0   (C) 2   (E) 4   (G) −4

(B) −10   (D) 8   (F) −8   (H) $\sqrt{3}$

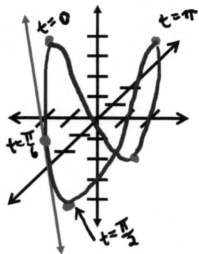

t=0   t=π

$t = \frac{\pi}{6}$

$t = \frac{\pi}{2}$

Equation of a line
Need (a) pt. on the line.
     (b) direction vector for
        the line

Tangent line

tangency pt. $(\sqrt{3}, 1, 2) \leftarrow r(t_0)$

$r'(t_0)$ to: parameter for
           tangency pt.

$r(t) = \langle 2\cos t, 2\sin t, 4\cos(2t) \rangle$
Find t_0 $\underbrace{\sqrt{3}}$ $\underbrace{1}$ $\underbrace{2}$

$2\cos(t_0) = \sqrt{3}$ $2\sin\frac{\pi}{6} = 2 \cdot \frac{1}{2} = 1 \checkmark$
$\cos(t_0) = \frac{\sqrt{3}}{2}$ $4\cos(\frac{\pi}{3}) = 4 \cdot \frac{1}{2} = 2 \checkmark$

$r'(t) = \langle -2\sin t, 2\cos t, -8\sin(2t) \rangle$ $t_0 = \frac{\pi}{6}$

$r'(\frac{\pi}{6}) = \langle -2\sin\frac{\pi}{6}, 2\cos\frac{\pi}{6}, -8\sin(\frac{\pi}{3}) \rangle = \langle -2 \cdot \frac{1}{2}, 2 \cdot \frac{\sqrt{3}}{2}, -8 \cdot \frac{\sqrt{3}}{2} \rangle$

$r'(\frac{\pi}{6}) = \langle -1, \sqrt{3}, -4\sqrt{3} \rangle$

Tangent line: $\left. \begin{array}{l} x = \sqrt{3} - t \\ y = 1 + \sqrt{3}t \\ z = 2 - 4\sqrt{3}t \end{array} \right\}$ intersects yz plane $\Rightarrow x = 0$ $\left\{ \begin{array}{l} x = 0 \\ y = 1 + \sqrt{3} \cdot \sqrt{3} = 1+3 = 4 \\ z = 2 - 4\sqrt{3}\sqrt{3} = 2 - 12 \end{array} \right.$
                                        $0 = \sqrt{3} - t$ z coordinate $\boxed{-10}$
                                          $\Rightarrow t = \sqrt{3}$

1.19) At what time t does the speed of the particle moving in space with its position function $\mathbf{r}(t) = \langle t^2, 3t, t^2 - 8t \rangle$ have its minimum value?

(A) $t = 0$

(B) $t = 1$

(C) $t = 2$

(D) $t = 3$

(E) $t = 4$

(F) $t = 6$

(G) $t = 8$

(H) None of these

1.19) At what time t does the speed of the particle moving in space with its position function $\mathbf{r}(t) = \langle t^2, 3t, t^2 - 8t \rangle$ have its minimum value?

(A) $t = 0$

(B) $t = 1$

(C) $t = 2$

(D) $t = 3$

(E) $t = 4$

(F) $t = 6$

(G) $t = 8$

(H) None of these

$$r(t) = \langle t^2, 3t, t^2 - 8t \rangle$$

$$r'(t) = \langle 2t, 3, 2t - 8 \rangle$$

$$Speed = |r'(t)|$$

the magnitude of the velocity vector

$$|r'(t)| = \sqrt{(2t)^2 + (3)^2 + (2t-8)^2}$$

$$|r'(t)| = \sqrt{4t^2 + 9 + 4t^2 - 32t + 64}$$

$$|r'(t)| = \sqrt{8t^2 - 32t + 73}$$

To minimize a square root function, you can minimize the function under the root.

Minimize $f = 8t^2 - 32t + 73 \leftarrow$ a parabola that opens upward. It is minimized at its vertex. This happens where $f' \stackrel{set}{=} 0$

$$f' = 16t - 32 \stackrel{set}{=} 0 \implies \boxed{t = 2}$$

1.20) Find the arclength of the curve

$$r(t) = \left\langle 1+t^2, 3+\frac{2}{3}t^3, 6-t \right\rangle$$

for $-3 \le t \le 3$.

(A) 20   (E) 24
(B) 21   (F) 12
(C) 42   (G) 16
(D) 53   (H) 37

1.20) Find the arclength of the curve

$$\mathbf{r}(t) = \left\langle 1 + t^2, 3 + \frac{2}{3}t^3, 6 - t \right\rangle$$

for $-3 \leq t \leq 3$.

(A) 20   (E) 24

(B) 21   (F) 12

(C) 42   (G) 16

(D) 53   (H) 37

$$\text{Arclength} = \int_{-3}^{3} |r'(t)|\, dt$$

$$r(t) = \langle 1 + t^2, 3 + \tfrac{2}{3}t^3, 6 - t \rangle$$

$$r'(t) = \langle 2t, 2t^2, -1 \rangle$$

$$|r'(t)| = \sqrt{4t^2 + 4t^4 + 1} = \sqrt{4t^4 + 4t^2 + 1}$$

$$|r'(t)| = \sqrt{(2t^2 + 1)^2} = 2t^2 + 1$$

$$\text{Arclength} = \int_{-3}^{3}(2t^2 + 1)\, dt = 2\int_{0}^{3}(2t^2 + 1)\, dt = 2\left[\frac{2t^3}{3} + t\right]_{0}^{3}$$

$$\underbrace{}_{\substack{\text{even} \\ \text{function}}}$$

$$= 2\left[\left(\tfrac{2}{3}(3)^3 + 3\right) - 0\right] = 2\left[\tfrac{2 \cdot 9}{18 + 3} + 3\right] = \boxed{42}$$

1.21) Calculate the arclength of the curve given parametrically by

$$x = 2t^2, \qquad y = \frac{8\sqrt{3}}{5}t^{5/2}, \qquad z = 2t^3$$

for $0 \le t \le 2$.

a) 144   b) 126   c) 92   d) 63   e) 24   f) 20   g) 5   h) 2

1.21) Calculate the arclength of the curve given parametrically by

$$x = 2t^2, \qquad y = \frac{8\sqrt{3}}{5}t^{5/2}, \qquad z = 2t^3$$

for $0 \le t \le 2$.

a) 144   b) 126   c) 92   d) 63   e) 24   f) 20   g) 5   h) 2

$$r(t) = \langle 2t^2, \tfrac{8\sqrt{3}}{5}t^{5/2}, 2t^3 \rangle$$

$$r'(t) = \langle 4t, \tfrac{8\sqrt{3}}{5} \cdot \tfrac{5}{2} t^{3/2}, 6t^2 \rangle = \langle 4t, 4\sqrt{3}\, t^{3/2}, 6t^2 \rangle$$

$$|r'(t)| = \sqrt{(4t)^2 + (4\sqrt{3}\, t^{3/2})^2 + (6t^2)^2} = \sqrt{16t^2 + 48t^3 + 36t^4}$$

$$|r'(t)| = \sqrt{36t^4 + 48t^3 + 16t^2} = \sqrt{4t^2(9t^2 + 12t + 4)}$$

$$|r'(t)| = \sqrt{4t^2(3t+2)^2} = 2|t(3t+2)| \quad \text{for } 0 \le t \le 2$$
$$\text{always positive}$$

$$|r'(t)| = 2t(3t+2) = 6t^2 + 4t$$

$$\text{Arclength} = \int_0^2 |r'(t)| \, dt = \int_0^2 (6t^2 + 4t) \, dt = \left[2t^3 + 2t^2 \right]_0^2$$

$$= 2 \cdot (2)^3 + 2(2)^2 - 0 = 16 + 8 = \boxed{24}$$

1.22) Find the arclength of the curve

$$\mathbf{r}(t) = \langle 2t, \ln t, t^2 \rangle$$

for $1 \le t \le e$.

a) e^4

b) $\sqrt{e}$

c) e

d) 1

e) e^2

f) $\dfrac{e}{2}$

1.22) Find the arclength of the curve

$$\mathbf{r}(t) = \langle 2t, \ln t, t^2 \rangle$$

for $1 \le t \le e$.

a) e^4 d) 1

b) $\sqrt{e}$ e) e^2

c) e f) $\dfrac{e}{2}$

$r(t) = \langle 2t, \ln t, t^2 \rangle$

$r'(t) = \langle 2, \frac{1}{t}, 2t \rangle$

$|r'(t)| = \sqrt{4 + \frac{1}{t^2} + 4t^2} = \sqrt{4t^2 + \frac{1}{t^2} + 4}$

$|r'(t)| = \sqrt{\frac{4t^4 + 4t^2 + 1}{t^2}} = \sqrt{\frac{(2t^2+1)^2}{t^2}}$

$|r'(t)| = \sqrt{\left(\frac{2t^2+1}{t}\right)^2} = \sqrt{\left(2t + \frac{1}{t}\right)^2} = 2t + \frac{1}{t}$

$\text{Arclength} = \int_1^e |r'(t)| \, dt = \int_1^e \left(2t + \frac{1}{t}\right) dt = \left[t^2 + \ln t\right]_1^e$

$= \left(e^2 + \ln e\right) - \left(1 + \ln 1\right) = e^2 + 1 - 1 + 0 = \boxed{e^2}$

1.23) Find the curvature of the curve below at $t = 1$

$\mathbf{r}(t) = \langle t^2, \ln t, t \ln t \rangle$.

(A) $\dfrac{2\sqrt{6}}{5}$

(B) $\sqrt{2}$

(C) $\dfrac{3\sqrt{2}}{16}$

(D) $\dfrac{2\sqrt{5}}{6}$

(E) $\dfrac{\sqrt{15}}{30}$

(F) $\dfrac{\sqrt{10}}{9}$

(G) $\dfrac{\sqrt{30}}{18}$

(H) 0

1.23) Find the curvature of the curve below at $t = 1$

$\mathbf{r}(t) = \langle t^2, \ln t, t \ln t \rangle.$

(A) $\dfrac{2\sqrt{6}}{5}$ (E) $\dfrac{\sqrt{15}}{30}$

(B) $\sqrt{2}$ (F) $\dfrac{\sqrt{10}}{9}$

(C) $\dfrac{3\sqrt{2}}{16}$ (G) $\dfrac{\sqrt{30}}{18}$

(D) $\dfrac{2\sqrt{5}}{6}$ (H) 0

$r(t) = \langle t^2, \ln t, t \cdot \ln t \rangle$

$r'(t) = \langle 2t, \frac{1}{t}, 1 \cdot \ln t + t \cdot \frac{1}{t} \rangle$

$r'(t) = \langle 2t, \frac{1}{t}, \ln t + 1 \rangle = v(t)$ velocity

$r''(t) = \langle 2, -\frac{1}{t^2}, \frac{1}{t} \rangle = a(t)$ acceleration

$\chi(1) = \dfrac{|v(1) \times a(1)|}{|v(1)|^3}$

$v(1) = \langle 2, 1, 1 \rangle \quad |v(1)| = \sqrt{4+1+1} = \sqrt{6}$

$a(1) = \langle 2, -1, 1 \rangle$

$v(1) \times a(1) = \begin{vmatrix} i & j & k \\ 2 & 1 & 1 \\ 2 & -1 & 1 \end{vmatrix} = \langle 1+1, -(2-2), -2-2 \rangle = \langle 2, 0, -4 \rangle$

$|v(1) \times a(1)| = \sqrt{4+0+16} = \sqrt{20} = 2\sqrt{5}$

$\chi(1) = \dfrac{2\sqrt{5}}{(\sqrt{6})^3} = \dfrac{2\sqrt{5}}{\sqrt{6}\sqrt{6}\sqrt{6}} = \dfrac{2\sqrt{5}}{6\sqrt{6}} = \dfrac{\sqrt{5}\sqrt{6}}{3\sqrt{6}\sqrt{6}} = \dfrac{\sqrt{30}}{3 \cdot 6} = \boxed{\dfrac{\sqrt{30}}{18}}$

1.24) Find the curvature of the curve below at $t = 0$

$$\mathbf{r}(t) = \frac{4}{9}(1+t)^{3/2}\,\mathbf{i} + \frac{4}{9}(1-t)^{3/2}\,\mathbf{j} + \frac{1}{3}t\mathbf{k}$$

a) 0   b) $\dfrac{4}{3}$ c) $\dfrac{2}{3}$ d) $\dfrac{\sqrt{2}}{3}$ e) $\dfrac{\sqrt{3}}{2}$ f) $\dfrac{1}{2}$ g) 1   h) 2

1.24) Find the curvature of the curve below at $t = 0$

$$\mathbf{r}(t) = \frac{4}{9}(1+t)^{3/2}\,\mathbf{i} + \frac{4}{9}(1-t)^{3/2}\,\mathbf{j} + \frac{1}{3}t\mathbf{k}$$

a) 0   b) $\frac{4}{3}$ c) $\frac{2}{3}$ d) $\frac{\sqrt{2}}{3}$ e) $\frac{\sqrt{3}}{2}$ f) $\frac{1}{2}$ g) 1   h) 2

$r'(t) = \left\langle \frac{4}{9}\cdot\frac{3}{2}(1+t)^{1/2}\cdot(1) , \frac{4}{9}\cdot\frac{3}{2}(1-t)^{1/2}(-1), \frac{1}{3}\right\rangle = \left\langle \frac{2}{3}\sqrt{1+t}, -\frac{2}{3}\sqrt{1-t}, \frac{1}{3}\right\rangle$

$r''(t) = \left\langle \frac{2}{3}\cdot\frac{1}{2\sqrt{1+t}}, -\frac{2}{3}\cdot\frac{1}{2\sqrt{1-t}}(-1), 0\right\rangle = \left\langle \frac{1}{3\sqrt{1+t}}, \frac{1}{3\sqrt{1-t}}, 0\right\rangle$

$r'(0) = \left\langle \frac{2}{3}, -\frac{2}{3}, \frac{1}{3}\right\rangle$ $|r'(0)| = \sqrt{\frac{4}{9} + \frac{4}{9} + \frac{1}{9}} = 1$

$r''(0) = \left\langle \frac{1}{3}, \frac{1}{3}, 0\right\rangle$

$r'(0) \times r''(0) = \begin{vmatrix} \mathbf{i} & \mathbf{j} & \mathbf{k} \\ \frac{2}{3} & -\frac{2}{3} & \frac{1}{3} \\ \frac{1}{3} & \frac{1}{3} & 0 \end{vmatrix} = \left\langle -\frac{1}{9}, -(0-\frac{1}{9}), \frac{2}{9}+\frac{2}{9}\right\rangle$

$r'(0) \times r''(0) = \left\langle -\frac{1}{9}, \frac{1}{9}, \frac{4}{9}\right\rangle$

$|r'(0) \times r''(0)| = \sqrt{\frac{1}{81} + \frac{1}{81} + \frac{16}{81}} = \sqrt{\frac{18}{81}} = \sqrt{\frac{2}{9}} = \frac{\sqrt{2}}{3}$

$K = \frac{|r'(0) \times r''(0)|}{|r'(0)|^3} = \frac{\frac{\sqrt{2}}{3}}{1^3} = \boxed{\frac{\sqrt{2}}{3}}$

1.25) Find the curvature of the curve below at $t = 0$

$$\mathbf{r}(t) = \left\langle \left(2\sqrt{2}\right)t, e^{2t}, e^{-2t} \right\rangle$$

(A) 1

(B) $\sqrt{2}$

(C) $2\sqrt{2}$

(D) $4\sqrt{2}$

(E) $\dfrac{\sqrt{2}}{2}$

(F) $\dfrac{\sqrt{2}}{4}$

(G) $\dfrac{\sqrt{2}}{8}$

(H) 0

1.25) Find the curvature of the curve below at $t = 0$

$$r(t) = \left\langle \left(2\sqrt{2}\right)t, e^{2t}, e^{-2t} \right\rangle$$

(A) 1

(E) $\dfrac{\sqrt{2}}{2}$

(B) $\sqrt{2}$

(F) $\dfrac{\sqrt{2}}{4}$

(C) $2\sqrt{2}$

(G) $\dfrac{\sqrt{2}}{8}$

(D) $4\sqrt{2}$

(H) 0

$v = \left\langle 2\sqrt{2}, 2e^{2t}, -2e^{-2t} \right\rangle$

$a = \left\langle 0, 4e^{2t}, 4e^{-2t} \right\rangle$

$v(0) = \left\langle 2\sqrt{2}, 2, -2 \right\rangle \qquad |v(0)| = \sqrt{8+4+4} = 4$

$a(0) = \left\langle 0, 4, 4 \right\rangle$

$v(0) \times a(0) = \begin{vmatrix} i & j & k \\ 2\sqrt{2} & 2 & -2 \\ 0 & 4 & 4 \end{vmatrix}$

$v(0) \times a(0) = \left\langle 16, -8\sqrt{2}, 8\sqrt{2} \right\rangle = 8\left\langle 2, -\sqrt{2}, \sqrt{2} \right\rangle$

$|v(0) \times a(0)| = 8 \cdot \sqrt{4+2+2} = 8\sqrt{8} = 16\sqrt{2}$

$K(0) = \dfrac{|v(0) \times a(0)|}{|v(0)|^3}$

$K(0) = \dfrac{16\sqrt{2}}{4^3} = \dfrac{16\sqrt{2}}{4\cdot4\cdot4}$

$K(0) = \boxed{\dfrac{\sqrt{2}}{4}}$

1.26) A ball is thrown eastward into the air from the origin (in the direction of the positive $x-$axis). The initial velocity is $50\mathbf{i} + 80\mathbf{k}$. The spin of the ball results in a southward acceleration of 4 ft/s, so the acceleration vector is $\mathbf{a} = -4\mathbf{j} - 32\mathbf{k}$. Where does the ball land and with what speed?

1.26) A ball is thrown eastward into the air from the origin $\left(\text{in the direction of the positive } x - \text{axis}\right)$. The initial velocity is $50\mathbf{i} + 80\mathbf{k}$. The spin of the ball results in a southward acceleration of 4 ft/s, so the acceleration vector is $\mathbf{a} = -4\mathbf{j} - 32\mathbf{k}$.

Where does the ball land and with what speed?

$$v(t) = \int a(t)\,dt = \int \langle 0, -4, -32 \rangle\,dt = \langle 0 + c_1, -4t + c_2, -32t + c_3 \rangle$$

$$v(t) = \langle 0, -4t, -32t \rangle + \langle c_1, c_2, c_3 \rangle$$

$$v(0) = \langle 0, 0, 0 \rangle + \langle c_1, c_2, c_3 \rangle \overset{\text{set}}{=} \langle 50, 0, 80 \rangle$$

$$v(t) = \langle 50, -4t, -32t + 80 \rangle$$

$$r(t) = \int v(t)\,dt = \int \langle 50, -4t, -32t + 80 \rangle\,dt = \langle 50t + c_4, -2t^2 + c_5, -16t^2 + 80t + c_6 \rangle$$

$$r(t) = \langle 50t, -2t^2, -16t^2 + 80t \rangle + \langle c_4, c_5, c_6 \rangle$$

$$r(0) = \langle 0, 0, 0 \rangle + \langle c_4, c_5, c_6 \rangle \overset{\text{set}}{=} \langle 0,0,0 \rangle \text{ since "from the origin"}$$

$$r(t) = \langle 50t, -2t^2, -16t^2 + 80t \rangle \text{ "hits the ground"} \Rightarrow z \text{ comp} = 0$$

$$-16t^2 + 80t = 0 \Rightarrow -16t(t - 5) = 0 \qquad t = 0 \quad t = 5$$

$$r(5) = \langle 50 \cdot 5, -2(5)^2, 0 \rangle = \langle 250, -2(25), 0 \rangle = \boxed{\langle 250, -50, 0 \rangle}$$

$$v(5) = \langle 50, -20, -32(5) + 80 \rangle = \langle 50, -20, -80 \rangle = 10\langle 5, -2, -8 \rangle$$

$$\text{Speed} = |v(5)| = 10\sqrt{25 + 4 + 64} = \boxed{10\sqrt{93}} \text{ ft/s}$$

1.27) A quarterback throws a football while standing at the very center of the field.

The ball leaves his hand at a height of 5 feet with an inital velocity of $\mathbf{v}_0 = 40\mathbf{i} + 35\mathbf{j} + 32\mathbf{k}$ ft/s.

Assume an acceleration of 32 ft/s² due to gravity and that the $\mathbf{i}$ vector points horizontally towards the right endzone and the $\mathbf{j}$ vector upward towards the sideline. See the figure below.

a. Determine the position function that gives the position of the ball t seconds after it is thrown.

b. The ball is caught at a height of 5 feet by a player standing vertically with both feet on the ground at the time that they caught the ball. Is the player in bounds or out of bounds when he receives the ball.

In bounds means on the playing field.

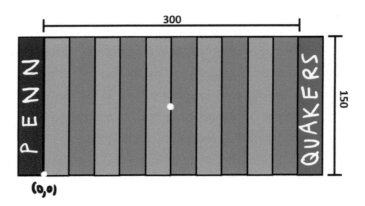

1.27) A quarterback throws a football while standing at the very center of the field. The ball leaves his hand at a height of 5 feet with an inital velocity of $v_0 = 40i + 35j + 32k$ ft/s. Assume an acceleration of 32 ft/s² due to gravity and that the i vector points horizontally towards the right endzone and the j vector upward towards the sideline. See the figure below.

a. Determine the position function that gives the position of the ball t seconds after it is thrown.

b. The ball is caught at a height of 5 feet by a player standing vertically with both feet on the ground at the time that they caught the ball. Is the player in bounds or out of bounds when he receives the ball. In bounds means on the playing field.

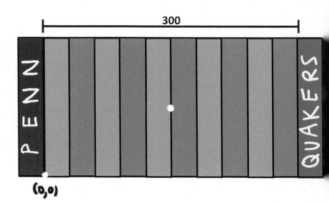

(0,0)

(a) $V(t) = \int a(t)dt = \int \langle 0, 0, -32 \rangle dt = \langle c_1, c_2, -32t + c_3 \rangle$

$V(t) = \langle 0, 0, -32t \rangle + \langle c_1, c_2, c_3 \rangle$ $V(0) = \langle c_1, c_2, c_3 \rangle$

$V(0) = \langle 40, 35, 32 \rangle$

$V(t) = \langle 40, 35, -32t + 32 \rangle$

$r(t) = \int V(t)dt = \int \langle 40, 35, -32t + 32 \rangle dt = \langle 40t, 35t, -16t^2 + 32t \rangle + \langle c_4, c_5, c_6 \rangle$

$r(0) = \langle 150, 75, 5 \rangle = \langle c_4, c_5, c_6 \rangle$ $\boxed{r(t) = \langle 40t + 150, 35t + 75, -16t^2 + 32t + 5 \rangle}$

   midfield   height
             off ground

(b) Find the time until the catch ⇒ set k comp = to 5.

$-16t^2 + 32t + 5 = 5 \Rightarrow -16t^2 + 32t = 0$

$-16t(t - 2) = 0$

$t = 0$ $t = 2 \rightarrow$ Find $r(2)$

$r(2) = \langle 80 + 150, 70 + 75, 5 \rangle = \langle 230, 145, 5 \rangle$

inbounds since i < 300 j < 150

1.28) A kid standing on top of a 40 m. cliff throws a ball upward at an angle of 60 with respect to the horizontal and with a speed of 20 m/s.

How high off the ground is the ball $\dfrac{4}{\sqrt{3}}$ seconds later?

Is it travelling upward or downward? USE $g = 10 \, m/s^2$

1.28) A kid standing on top of a 40 *m.* cliff throws a ball upward
at an angle of 60 with respect to the horizontal and with a speed of 20 m/s.
How high off the ground is the ball $\dfrac{4}{\sqrt{3}}$ seconds later?
Is it travelling upward or downward? USE $g = 10 \, m/s^2$

$$\mathbf{r}(t) = \left\langle \left(|\mathbf{v}_0|\cos\theta\right)t, h + \left(|\mathbf{v}_0|\sin\theta\right)t - \frac{1}{2}gt^2 \right\rangle$$

$h = 40 \qquad \theta = 60° \qquad |V_0| = 20 \qquad g = 10$

$\qquad\qquad \cos\theta = \frac{1}{2} \qquad \sin\theta = \sqrt{3}/2$

$r(t) = \left\langle 20 \cdot \frac{1}{2} t, \; 40 + 20 \cdot \frac{\sqrt{3}}{2} t - \frac{1}{2} \cdot 10t^2 \right\rangle$

$r(t) = \left\langle 10t, \; \underbrace{40 + 10\sqrt{3}t - 5t^2}_{\substack{\text{how high off the} \\ \text{ground} \Rightarrow j \text{ comp.}}} \right\rangle$

$40 + 10\sqrt{3} \cdot \dfrac{4}{\sqrt{3}} - 5\left(\dfrac{4}{\sqrt{3}}\right)^2 = 40 + 40 - 5\left(\dfrac{16}{3}\right)$

$\qquad = 80 - \dfrac{80}{3} = 80\left(1 - \frac{1}{3}\right) = 80 \cdot \dfrac{2}{3} = \boxed{\dfrac{160}{3} \, m}$

The j component of the velocity vector determines whether
the ball is moving upward (+) or downward (-)

$r'(t) = \left\langle 10, \; \underbrace{10\sqrt{3} - 10t} \right\rangle$

$\underset{@ \frac{4}{\sqrt{3}}}{j\text{-comp}} : \quad 10\sqrt{3} - 10 \cdot \dfrac{4}{\sqrt{3}} \dfrac{\sqrt{3}}{\sqrt{3}} = 10\sqrt{3} - \dfrac{40}{3}\sqrt{3}$

$\qquad\qquad\qquad\qquad = \underbrace{(10 - 13.3)}\sqrt{3}$

$\qquad\qquad\qquad\qquad \text{negative} \Rightarrow \boxed{\text{downward}}$

1.29) For the position function below, write the acceleration

vector **a** in the form $\mathbf{a} = a_T \mathbf{T} + a_N \mathbf{N}$ at $t = \dfrac{\pi}{2}$ without finding **T** and **N**.

$r(t) = (3\sin t)\mathbf{i} + (2\cos t)\mathbf{j} + (-\sin 2t)\mathbf{k}$

1.29) For the position function below, write the acceleration

vector **a** in the form $\mathbf{a} = a_T\mathbf{T} + a_N\mathbf{N}$ at $t = \dfrac{\pi}{2}$ without finding **T** and **N**.

$$r(t) = (3\sin t)\mathbf{i} + (2\cos t)\mathbf{j} + (-\sin 2t)\mathbf{k}$$

$a_T = \dfrac{v \cdot a}{|v|}$

$a_N = \dfrac{|v \times a|}{|v|}$

$v = \langle 3\cos t, -2\sin t, -2\cos 2t \rangle$

$a = \langle -3\sin t, -2\cos t, 4\sin(2t) \rangle$

$\sin \frac{\pi}{2} = 1$

$\cos \frac{\pi}{2} = 0$

$\frac{2\pi}{2} = \pi$

$\cos \pi = -1$

$\sin \pi = 0$

$v(\pi/2) = \langle 0, -2, 2 \rangle \quad |v| = 2\langle 0, -1, 1 \rangle$

$a(\pi/2) = \langle -3, 0, 0 \rangle \quad |v| = 2\sqrt{2}$

$\boxed{v \cdot a = 0} \Rightarrow \boxed{a_T = 0}$

$a_N:$

$v \times a = \begin{vmatrix} i & j & k \\ 0 & -2 & 2 \\ -3 & 0 & 0 \end{vmatrix} = \langle 0, -(0+6), 0-6 \rangle = \langle 0, -6, -6 \rangle$

$|v \times a| = -6\langle 0, 1, 1 \rangle = \boxed{6\sqrt{2} = |v \times a|}$

$\underbrace{|-6|\sqrt{2}}$

$\dfrac{|v \times a|}{|v|} = \dfrac{6\sqrt{2}}{2\sqrt{2}} = 3 \qquad \boxed{a_N = 3}$

$$\boxed{a = 0 \cdot T + 3N}$$

1.30) For the position function below, write the acceleration vector $\mathbf{a}$ in the form $\mathbf{a} = a_T \mathbf{T} + a_N \mathbf{N}$ at $t = 0$ without finding $\mathbf{T}$ and $\mathbf{N}$.

$$\mathbf{r}(t) = \left(2 + 3t + 3t^2\right)\mathbf{i} + \left(4t + 4t^2\right)\mathbf{j} - \left(6\cos t\right)\mathbf{k}$$

1.30) For the position function below, write the acceleration
vector **a** in the form $\mathbf{a} = a_T\mathbf{T} + a_N\mathbf{N}$ at $t = 0$ without finding **T** and **N**.
$$\mathbf{r}(t) = (2+3t+3t^2)\mathbf{i} + (4t+4t^2)\mathbf{j} - (6\cos t)\mathbf{k}$$

$\mathbf{r}(t) = (2+3t+3t^2)\mathbf{i} + (4t+4t^2)\mathbf{j} - (6\cos t)\mathbf{k}$

$r'(t) = \langle 3+6t, 4+8t, 6\sin t \rangle$ $r'(0) = \langle 3,4,0 \rangle = v$

$r''(t) = \langle 6, 8, 6\cos t \rangle$ $r''(0) = \langle 6,8,6 \rangle = a$

$a_T = \dfrac{v \cdot a}{|v|} = \dfrac{50}{5} = 10$ $v \cdot a = 18+32 = 50$

$|v| = \sqrt{9+16} = 5$

$a_N = \dfrac{|v \times a|}{|v|} = \dfrac{30}{5} = 6$ $v \times a = \begin{vmatrix} i & j & k \\ 3 & 4 & 0 \\ 6 & 8 & 6 \end{vmatrix} = \langle 24, -18, 24-24 \rangle$

$\boxed{a = 10T + 6N}$ $= \langle 24, -18, 0 \rangle$

$v \times a = 6\langle 4, -3, 0 \rangle$

$|v \times a| = 6\sqrt{16+9} = 6 \cdot 5 = 30$

1.31) Find the torsion of the curve below at $t = 0$.

$$r(t) = \left\langle e^t \sin t, e^t \cos t, e^t \right\rangle$$

a) $\dfrac{2}{\sqrt{6}}$　　b) $\dfrac{-2}{\sqrt{6}}$　　c) $\dfrac{1}{3}$　　d) $\dfrac{-1}{3\sqrt{6}}$　　e) $\dfrac{-1}{3}$　　f) None of these

1.31) Find the torsion of the curve below at $t = 0$.

$$\mathbf{r}(t) = \left\langle e^t \sin t, e^t \cos t, e^t \right\rangle$$

a) $\dfrac{2}{\sqrt{6}}$ b) $\dfrac{-2}{\sqrt{6}}$ c) $\dfrac{1}{3}$ d) $\dfrac{-1}{3\sqrt{6}}$ e) $\dfrac{-1}{3}$ f) None of these

$$\mathbf{r}(t) = \left\langle e^t \sin t, e^t \cos t, e^t \right\rangle$$

$$\mathbf{r}'(t) = \left\langle e^t \sin t + e^t \cos t, e^t \cos t - e^t \sin t, e^t \right\rangle$$

$$\mathbf{r}'(t) = \left\langle e^t(\sin t + \cos t), e^t(\cos t - \sin t), e^t \right\rangle \quad \mathbf{r}'(0) = \langle 1, 1, 1 \rangle$$

$$\mathbf{r}''(t) = \left\langle e^t(\sin t + \cos t) + e^t(\cos t - \sin t), e^t(\cos t - \sin t) + e^t(-\sin t - \cos t), e^t \right\rangle$$

$$\mathbf{r}''(t) = \left\langle 2e^t \cos t, -2e^t \sin t, e^t \right\rangle \qquad \mathbf{r}''(0) = \langle 2, 0, 1 \rangle$$

$$\mathbf{r}'''(t) = \left\langle 2e^t \cos t - 2e^t \sin t, -2e^t \sin t - 2e^t \cos t, e^t \right\rangle$$

$$\mathbf{r}'''(t) = \left\langle 2e^t(\cos t - \sin t), -2e^t(\sin t + \cos t), e^t \right\rangle \quad \mathbf{r}'''(0) = \langle 2, -2, 1 \rangle$$

$$\mathbf{r}' \cdot (\mathbf{r}'' \times \mathbf{r}''') = \begin{vmatrix} 1 & 1 & 1 \\ 2 & 0 & 1 \\ 2 & -2 & 1 \end{vmatrix} = 1 \cdot (0+2) - 1(2-2) + 1(-4)$$

$$= 2 - 0 - 4 = -2$$

$$\mathbf{r}' \times \mathbf{r}'' = \begin{vmatrix} i & j & K \\ 1 & 1 & 1 \\ 2 & 0 & 1 \end{vmatrix} = \langle 1, -(1-2), 0-2 \rangle$$

$$= \langle 1, 1, -2 \rangle \qquad |\mathbf{r}' \times \mathbf{r}''| = \sqrt{1+1+4}$$

$$= \sqrt{6}$$

$$\tau = \frac{\mathbf{r}' \cdot (\mathbf{r}'' \times \mathbf{r}''')}{|\mathbf{r}' \times \mathbf{r}''|^2} = \frac{-2}{6} = \boxed{\frac{-1}{3}}$$

1.32) Find the torsion of the curve below at $t = 0$.

$$\mathbf{r}(t) = \frac{4}{9}(1+t)^{3/2}\,\mathbf{i} + \frac{4}{9}(1-t)^{3/2}\,\mathbf{j} + \frac{1}{3}t\,\mathbf{k}$$

a) $\dfrac{-4}{9}$ b) $\dfrac{1}{6}$ c) $\dfrac{1}{3}$ d) $\dfrac{1}{27}$ e) $\dfrac{1}{9}$ f) None of these

1.32) Find the torsion of the curve below at $t = 0$.

$$\mathbf{r}(t) = \frac{4}{9}(1+t)^{3/2}\,\mathbf{i} + \frac{4}{9}(1-t)^{3/2}\,\mathbf{j} + \frac{1}{3}t\,\mathbf{k}$$

a) $\frac{-4}{9}$ b) $\frac{1}{6}$ c) $\frac{1}{3}$ d) $\frac{1}{27}$ e) $\frac{1}{9}$ f) None of these

$$r(t) = \left\langle \tfrac{4}{9}(1+t)^{3/2},\ \tfrac{4}{9}(1-t)^{3/2},\ \tfrac{1}{3}t \right\rangle$$

$$r'(t) = \left\langle \tfrac{4}{9}\tfrac{3}{2}(1+t)^{1/2},\ \tfrac{4}{9}\tfrac{3}{2}(1-t)^{1/2}(-1),\ \tfrac{1}{3} \right\rangle \quad \text{simplify}$$

$$\bullet\ r'(t) = \left\langle \tfrac{2}{3}(1+t)^{1/2},\ -\tfrac{2}{3}(1-t)^{1/2},\ \tfrac{1}{3} \right\rangle \implies \boxed{r'(0) = \left\langle \tfrac{2}{3}, -\tfrac{2}{3}, \tfrac{1}{3} \right\rangle}$$

$$r''(t) = \left\langle \tfrac{2}{3}\tfrac{1}{2}(1+t)^{-1/2},\ -\tfrac{2}{3}\tfrac{1}{2}(1-t)^{-1/2}(-1),\ 0 \right\rangle \quad \text{simplify}$$

$$\bullet\ r''(t) = \left\langle \tfrac{1}{3}(1+t)^{-1/2},\ \tfrac{1}{3}(1-t)^{-1/2},\ 0 \right\rangle \implies \boxed{r''(0) = \left\langle \tfrac{1}{3}, \tfrac{1}{3}, 0 \right\rangle}$$

$$r'''(t) = \left\langle \tfrac{1}{3}\tfrac{-1}{2}(1+t)^{-3/2},\ \tfrac{1}{3}\left(\tfrac{-1}{2}\right)(1-t)^{-3/2}(-1),\ 0 \right\rangle$$

$$\bullet\ r'''(t) = \left\langle -\tfrac{1}{6}(1+t)^{-3/2},\ \tfrac{1}{6}(1-t)^{-3/2},\ 0 \right\rangle \implies \boxed{r'''(0) = \left\langle -\tfrac{1}{6}, \tfrac{1}{6}, 0 \right\rangle}$$

$$r' \cdot (r'' \times r''') = \begin{vmatrix} \tfrac{2}{3} & -\tfrac{2}{3} & \tfrac{1}{3} \\ \tfrac{1}{3} & \tfrac{1}{3} & 0 \\ -\tfrac{1}{6} & \tfrac{1}{6} & 0 \end{vmatrix} = \tfrac{1}{3}\begin{vmatrix} \tfrac{1}{3} & \tfrac{1}{3} \\ -\tfrac{1}{6} & \tfrac{1}{6} \end{vmatrix} = \tfrac{1}{3}\left(\tfrac{1}{18} + \tfrac{1}{18}\right) = \tfrac{1}{3}\cdot\tfrac{2}{18} = \boxed{\tfrac{1}{27}}$$

$$r' \times r'' = \begin{vmatrix} i & j & k \\ \tfrac{2}{3} & -\tfrac{2}{3} & \tfrac{1}{3} \\ \tfrac{1}{3} & \tfrac{1}{3} & 0 \end{vmatrix} = \left\langle 0 - \tfrac{1}{9},\ -\left(0 - \tfrac{1}{9}\right),\ \tfrac{2}{9} + \tfrac{2}{9} \right\rangle = \left\langle -\tfrac{1}{9}, \tfrac{1}{9}, \tfrac{4}{9} \right\rangle$$

$$|r' \times r''| = \tfrac{1}{9}\sqrt{1 + 1 + 16} = \tfrac{\sqrt{18}}{9} = \tfrac{3\sqrt{2}}{9} = \tfrac{\sqrt{2}}{3}$$

$$|r' \times r''|^2 = \boxed{\tfrac{2}{9}}$$

$$T(0) = \frac{r'\cdot(r''\times r''')}{|r'\times r''|^2} = \frac{1/27}{2/9} = \tfrac{1}{27}\cdot\tfrac{9}{2} = \boxed{\tfrac{1}{6}}$$

1.33) Find the maximum torsion of the curve

$$\mathbf{r}(t) = t\,\mathbf{i} + \frac{t^2}{2}\mathbf{j} + \frac{t^3}{3}\,\mathbf{k}$$

A) $\dfrac{1}{2}$ B) 1   C) 2   D) $\dfrac{1}{4}$

E) $\dfrac{3}{2}$ F) $\dfrac{2}{3}$ G) $\dfrac{3}{8}$ H) None of these

1.33) Find the maximum torsion of the curve

$$\mathbf{r}(t) = t\,\mathbf{i} + \frac{t^2}{2}\,\mathbf{j} + \frac{t^3}{3}\,\mathbf{k}$$

A) $\dfrac{1}{2}$ B) 1 C) 2 D) $\dfrac{1}{4}$

E) $\dfrac{3}{2}$ F) $\dfrac{2}{3}$ G) $\dfrac{3}{8}$ H) None of these

$$\left.\begin{array}{l} r = \langle t, \frac{t^2}{2}, \frac{t^3}{3}\rangle \\[4pt] r' = \langle 1, t, t^2\rangle \\[4pt] r'' = \langle 0, 1, 2t\rangle \\[4pt] r''' = \langle 0, 0, 2\rangle \end{array}\right\} \quad r'\cdot(r'\times r''') = \begin{vmatrix} 1 & t & t^2 \\ 0 & 1 & 2t \\ 0 & 0 & 2 \end{vmatrix} = 1\begin{vmatrix} 1 & 2t \\ 0 & 2 \end{vmatrix} + 0 + 0 = 2$$

$$r'\times r'' = \begin{vmatrix} i & j & k \\ 1 & t & t^2 \\ 0 & 1 & 2t \end{vmatrix} = \langle 2t^2 - t^2, -(2t-0), 1-0\rangle$$

$$r'\times r'' = \langle t^2, -2t, 1\rangle$$

$$\tau = \frac{r'\cdot(r''\times r''')}{|r'\times r''|^2}$$

$$|r'\times r''| = \sqrt{(t^2)^2 + (-2t)^2 + (1)^2} = \sqrt{t^4 + 4t^2 + 1}$$

$$\tau = \frac{2}{t^4 + 4t^2 + 1}$$

$\tau(t)$ is an even function, τ is always positive

as $t \to \infty$ $\tau \to 0$

& $t \to -\infty$ $\tau \to 0$

$$\tau = 2(t^4 + 4t^2 + 1)^{-1}$$

$$\tau' = 2\cdot(-1)(t^4 + 4t^2 + 1)^{-2}\cdot(4t^3 + 8t)$$

$$\tau' = \frac{-2(4t^3 + 8t)}{(t^4 + 4t^2 + 1)^2} \overset{\text{set}}{=} 0 \;\Rightarrow\; -8t(t^2 + 2) = 0$$

$$-8t = 0 \qquad t^2 + 2 = 0$$

$$t = 0 \qquad t^2 = -2 \;\; \text{not real}$$

$$\tau \xrightarrow{\;+\,+\,+\,\hat{\;}\,-\,-\,-\;}$$

$$\underset{\text{local max}}{\text{↑}}$$

actually abs max.

$$\tau(0) = \frac{2}{0 + 0 + 1} = \boxed{2}$$

Section 2: Multivariable Calculus

2.1) Find and sketch the domain of $f(x,y) = \dfrac{\ln(y-x)}{\sqrt{x-y+4}}$

2.1) Find and sketch the domain of $f(x,y) = \dfrac{\ln(y-x)}{\sqrt{x-y+4}}$

Input to $\ln$ function must be greater than 0

$y - x > 0 \implies y > x$

Input to $\sqrt{\ }$ function must be greater than or $=$ to 0, but since it is in the denom, it can't be $=$ too.

$x - y + 4 > 0 \implies -y > -x-4 \implies y < x+4$

2.2) Let $f(x,y) = e^{xy} \ln y$. Find $f_y\left(e, \dfrac{1}{e}\right) - f_x\left(e, \dfrac{1}{e}\right)$.

A) 1   B) 2   C) e^2 D) $2e^2$

E) 0   F) -1 G) e H) None of these

2.2) Let $f(x,y) = e^{xy} \ln y$. Find $f_y\left(e, \dfrac{1}{e}\right) - f_x\left(e, \dfrac{1}{e}\right)$.

A) 1   B) 2   C) e^2 D) $2e^2$

E) 0   F) -1 G) e H) None of these

$$f(x,y) = e^{xy} \ln y$$

f_x = the partial deriv. w.r.t. x, hold y constant

$$f_x = (e^{xy} \cdot y) \ln y + e^{xy} \cdot 0 = y e^{xy} \ln y$$

$$f_x(e, \tfrac{1}{e}) = \tfrac{1}{e} e^1 \cdot \ln(\tfrac{1}{e}) = \ln 1 - \ln e = 0 - 1 = -1$$

f_y = the partial deriv. w.r.t. y, hold x constant

$$f_y = (e^{xy} \cdot x) \ln y + e^{xy} \cdot \tfrac{1}{y}$$

$$f_y(e, \tfrac{1}{e}) = e \cdot e \cdot \ln(\tfrac{1}{e}) + e \cdot \tfrac{1}{\tfrac{1}{e}} = e^2 [\ln 1 - \ln e] + e^2$$

$$= -e^2 + e^2 = 0$$

$$f_y(e, \tfrac{1}{e}) - f_x(e, \tfrac{1}{e}) = 0 - (-1) = \boxed{1}$$

2.3) Let $f(x,y) = \sqrt{\dfrac{x}{y}}$. Find   a) $f_x(9,4)$ and   b) $f_y(9,4)$.

2.3) Let $f(x,y) = \sqrt{\dfrac{x}{y}}$. Find a) $f_x(9,4)$ and b) $f_y(9,4)$.

$$f(x,y) = \sqrt{\frac{x}{y}} = \frac{\sqrt{x}}{\sqrt{y}} = \frac{1}{\sqrt{y}} \cdot \sqrt{x} = x^{1/2} \cdot y^{-1/2}$$

$$f_x(x,y) = \frac{1}{\sqrt{y}} \cdot \frac{1}{2} x^{-1/2} = \frac{1}{\sqrt{y}} \cdot \frac{1}{2\sqrt{x}} = \frac{1}{2\sqrt{xy}}$$
Hold y constant

$$f_y(x,y) = \sqrt{x} \cdot \frac{-1}{2} y^{-3/2} = \frac{-\sqrt{x}}{2y^{3/2}} = \frac{-\sqrt{x}}{2y\sqrt{y}}$$
Hold x constant

$$f_x(9,4) = \frac{1}{2\sqrt{9\cdot4}} = \frac{1}{2\sqrt{36}} = \frac{1}{2\cdot6} = \boxed{\frac{1}{12}}$$

$$f_y(9,4) = \frac{-\sqrt{9}}{2\cdot4\sqrt{4}} = \frac{-3}{2\cdot4\cdot2} = \boxed{\frac{-3}{16}}$$

2.4) Let $g(x,y) = x\sin(x^2 y)$. Find $N = g_{xy}\left(\sqrt{3}, \dfrac{\pi}{2}\right)$.

A) 0   B) 2π C) 3π D) 4π

E) 6π F) 8π G) 9π H) None of these

2.4) Let $g(x, y) = x \sin(x^2 y)$. Find $N = g_{xy}\left(\sqrt{3}, \dfrac{\pi}{2}\right)$.

A) 0   B) 2π C) 3π D) 4π

E) 6π F) 8π G) 9π H) None of these

$g(x,y) = x \sin(x^2 y)$

$g_x = 1 \cdot \sin(x^2 y) + x \cos(x^2 y) \cdot 2xy$

$g_x = \boxed{\sin(x^2 y)} + \boxed{2x^2 y \cos(x^2 y)}$

_____

$g_{xy} = $ mixed partial deriv., the y partial deriv. of g_x

$g_{xy} = \boxed{\cos(x^2 y) \cdot x^2} + \underbrace{2x^2}_{y\,deriv.} \cos(x^2 y) + 2x^2 y \cdot \underbrace{(-\sin(x^2 y) \cdot x^2)}_{y\,deriv.}$ simplify

$g_{xy} = \underbrace{x^2 \cos(x^2 y)}_{} + 2x^2 \cos(x^2 y) - 2x^4 y \sin(x^2 y)$ combine

$g_{xy} = 3x^2 \cos(x^2 y) - 2x^4 y \sin(x^2 y)$

$g_{xy}\left(\sqrt{3}, \frac{\pi}{2}\right) = 3 \cdot 3 \cdot \underbrace{\cos\left(\frac{3\pi}{2}\right)}_{0} - 2 \cdot 9 \cdot \frac{\pi}{2} \sin\left(\frac{3\pi}{2}\right) = -9\pi(-1) = \boxed{9\pi}$

2.5) Let $w = x^2 + \dfrac{y}{x}$ and $x = u - 2v + 1, y = 2u + v - 2$.

Find $\dfrac{\partial w}{\partial v}$ when $u = v = 0$.

a) −9   b) −8   c) −7   d) −6   e) −4   f) 0   g) 4   h) 8

2.5) Let $w = x^2 + \dfrac{y}{x}$ and $x = u - 2v + 1$, $y = 2u + v - 2$.

Find $\dfrac{\partial w}{\partial v}$ when $u = v = 0$.

a) −9   b) −8   c) −7   d) −6   e) −4   f) 0   g) 4   h) 8

$w(x,y) = x^2 + \dfrac{y}{x}$

$\dfrac{\partial w}{\partial v} = \dfrac{\partial w}{\partial x} \cdot \dfrac{\partial x}{\partial v} + \dfrac{\partial w}{\partial y} \cdot \dfrac{\partial y}{\partial v} = 4(-2) + 1(1)$

at $u = v = 0$

$= -8 + 1 = \boxed{-7}$

$x(u,v) = u - 2v + 1$
$y(u,v) = 2u + v - 2$

$\dfrac{\partial w}{\partial x} = 2x + y \cdot \dfrac{-1}{x^2}$

Hold y constant
_____

$\dfrac{\partial w}{\partial y} = 0 + \dfrac{1}{x} \cdot 1$

Hold x constant

$\dfrac{\partial x}{\partial v} = -2$

Hold u constant
_____

$\dfrac{\partial y}{\partial v} = 1$

Hold u constant

$u = v = 0$

$\Rightarrow x = 1$
$\Rightarrow y = -2$
_____

$\dfrac{\partial w}{\partial x}(1, -2) = 2 + 2 = 4$

$\dfrac{\partial w}{\partial y}(1, -2) = 1$

2.6) Let be a function of x and y. Find $\dfrac{\partial z}{\partial x}$ at $\left(\sqrt{e},0,1\right)$ if

$$\ln\left(x^2 + y^2\right) + x\ln z = \cos\left(xyz\right)$$

2.6) Let be a function of x and y. Find $\dfrac{\partial z}{\partial x}$ at $\left(\sqrt{e},0,1\right)$ if

$$\ln\left(x^2+y^2\right)+x\ln z = \cos\left(xyz\right)$$

z is implicitly defined as a function of x and y

$\dfrac{\partial z}{\partial x}$ will be found by first creating a function $F(x,y,z)=0$

and then finding $-\dfrac{F_x}{F_z}$ at $\left(\sqrt{e},0,1\right)$

$\ln(x^2+y^2) + x\ln z = \cos(xyz)$

$F = \ln(x^2+y^2) + x\ln z - \cos(xyz)$

$F_x = \dfrac{1}{x^2+y^2}(2x) + \ln z - \left(-\sin(xyz)\right)(yz) = \dfrac{2x}{x^2+y^2} + yz\,\ln z\,\sin(xyz)$

$F_z = x\cdot\dfrac{1}{z} - \left(-\sin(xyz)\right)xy = \dfrac{x}{z} + xy\,\sin(xyz)$

$\dfrac{\partial z}{\partial x}\Big|_{(\sqrt{e},0,1)} = -\dfrac{F_x(\sqrt{e},0,1)}{F_z(\sqrt{e},0,1)} = -\dfrac{\left(\dfrac{2\sqrt{e}}{e}+0\right)}{\sqrt{e}+0} = \dfrac{\dfrac{-2}{\sqrt{e}}}{\sqrt{e}} = \boxed{\dfrac{-2}{e}}$

2.7) Find the derivative of $f(x, y, z) = xyz$ in the direction of the velocity vector of the helix $\mathbf{r}(t) = \langle \cos(6t), \sin(6t), 6t \rangle$ at $t = \dfrac{\pi}{6}$.

2.7) Find the derivative of $f(x, y, z) = xyz$ in the direction of the velocity

vector of the helix $\mathbf{r}(t) = \langle \cos(6t), \sin(6t), 6t \rangle$ at $t = \dfrac{\pi}{6}$.

$f(x,y,z) = xyz$ in the direction of the velocity vector of

$\mathbf{r}(t) = \langle x(t), y(t), z(t) \rangle$ $\mathbf{r} = \langle \cos(6t), \sin(6t), 6t \rangle$ at $t = \frac{\pi}{6}$

$\mathbf{r}(\frac{\pi}{6}) = \langle \cos\pi, \sin\pi, \pi \rangle$ $\mathbf{r}'(t) = \langle -6\sin(6t), 6\cos(6t), 6 \rangle$

$\mathbf{r}(\pi/6) = \langle -1, 0, \pi \rangle$ $\mathbf{r}'(\frac{\pi}{6}) = \langle -6\sin\pi, 6\cos\pi, 6 \rangle = \langle 0, -6, 6 \rangle$

$x = -1 \quad y = 0 \quad z = \pi$ $\mathbf{v} = \langle 0, -6, 6 \rangle \quad |\mathbf{v}| = 6\sqrt{2} \quad \mathbf{u} = \frac{6\langle 0, -1, 1 \rangle}{6\sqrt{2}} = \langle 0, \frac{-1}{\sqrt{2}}, \frac{1}{\sqrt{2}} \rangle$

when $t = \pi/6$ $\mathbf{v} = 6\langle 0, -1, 1 \rangle$

$D_u f(-1, 0, \pi) = \nabla f(-1, 0, \pi) \cdot \mathbf{u} = \langle 0, -\pi, 0 \rangle \cdot \langle 0, \frac{-1}{\sqrt{2}}, \frac{1}{\sqrt{2}} \rangle = \boxed{\dfrac{\pi}{\sqrt{2}} \text{ or } \dfrac{\sqrt{2}\pi}{2}}$

$f_x = yz \qquad f_x(-1, 0, \pi) = 0$

$f_y = xz \qquad f_y(-1, 0, \pi) = -\pi$

$f_z = xy \qquad f_z(-1, 0, \pi) = 0$

2.8) Find the derivative of the function

$$f(x,y,z) = x^2y + x\sqrt{1+z}$$

at $(1,2,3)$ in the direction of $\mathbf{v} = 2\mathbf{i} + \mathbf{j} - 2\mathbf{k}$.

a) 2   b) 4   c) $\dfrac{25}{6}$ d) $\dfrac{25}{3}$ e) $\dfrac{25}{2}$ f) $\dfrac{9}{2}$ g) $\dfrac{13}{3}$ h) 0

2.8) Find the derivative of the function

$$f(x,y,z) = x^2 y + x\sqrt{1+z}$$

at $(1,2,3)$ in the direction of $\mathbf{v} = 2\mathbf{i} + \mathbf{j} - 2\mathbf{k}$.

a) 2 b) 4 c) $\dfrac{25}{6}$ d) $\dfrac{25}{3}$ e) $\dfrac{25}{2}$ f) $\dfrac{9}{2}$ g) $\dfrac{13}{3}$ h) 0

$f(x,y,z) = x^2 y + x\sqrt{1+z}$ at $(1,2,3)$ in the $\mathbf{v} = \langle 2,1,-2 \rangle$ direction

Directional $D_u f(x_0, y_0) = \nabla f(x_0, y_0, z_0) \cdot u$
Derivative

$\nabla f(x_0, y_0, z_0) = \langle f_x(x_0, y_0, z_0), f_y(x_0, y_0, z_0), f_z(x_0, y_0, z_0) \rangle$

$D_u f(1,2,3) = \langle 6, 1, \tfrac{1}{4} \rangle \cdot \tfrac{1}{3}\langle 2,1,-2 \rangle = \tfrac{1}{3}(12 + 1 - \tfrac{1}{2}) = \tfrac{1}{3} \cdot \tfrac{25}{2} = \boxed{\dfrac{25}{6}}$

$f_x = 2xy + \sqrt{1+z}$ $f_x(1,2,3) = 2 \cdot 1 \cdot 2 + \sqrt{1+3} = 4+2 = 6$ $|V| = \sqrt{4+1+4} = \sqrt{9} = 3$

$f_y = x^2$ $f_y(1,2,3) = 1$ $u = \dfrac{V}{|V|} = \tfrac{1}{3}\langle 2,1,-2 \rangle$

$f_z = \dfrac{x}{2\sqrt{1+z}}$ $f_z(1,2,3) = \dfrac{1}{2\sqrt{4}} = \dfrac{1}{4}$

2.9) Find the equation of the tangent plane to the surface $z = \dfrac{x}{\sqrt{y}}$ when $x = y = 4$.

The z coordinate where the tangent plane intersects the z − axis is

a) −2   b) −1   c) 0   d) 1   e) 2   f) 3   g) 4   h) 5

2.9) Find the equation of the tangent plane to the surface $z = \dfrac{x}{\sqrt{y}}$ when $x = y = 4$.

The z coordinate where the tangent plane intersects the z – axis is

a) −2   b) −1   c) 0   d) 1   e) 2   f) 3   g) 4   h) 5

$$z = \frac{x}{\sqrt{y}}$$
at $(4,4)$

$z = f(x,y)$ is explicitly defined

$\Downarrow$

tangent plane to $f(x,y)$ at (x_0, y_0)

$z - f(x_0, y_0) = f_x(x_0, y_0)(x - x_0) + f_y(x_0, y_0)(y - y_0)$

$z - 2 = \frac{1}{2}(x - 4) - \frac{1}{4}(y - 4)$ mult. by 4

$4z - 8 = 2x - 8 - y + 4 \Rightarrow \boxed{2x - y - 4z = -4}$

intersect z-axis $\Rightarrow x = y = 0$ $-4z = -4 \Rightarrow \boxed{z = 1}$

$f(4,4) = \frac{4}{\sqrt{4}} = \frac{4}{2} = 2$

$f_x = \frac{1}{\sqrt{y}} \cdot 1$ $f_x(4,4) = \frac{1}{\sqrt{4}} = \frac{1}{2}$

$f_y = x \cdot \frac{-1}{2} y^{-3/2} = \frac{-x}{2 y^{3/2}}$ $f_y(4,4) = \frac{-4}{2 \cdot 4^{3/2}} = \frac{-4}{2 \cdot 8} = \frac{-4}{16} = \frac{-1}{4}$

think of f as $f = \frac{1}{\sqrt{y}} x$ think of f as $f = x \cdot y^{-1/2}$
hold y constant, take deriv.   hold x constant, take deriv.
with respect to x with respect to y

2.10) Find the equation of the tangent plane to the surface

$$x^2 + y^2 + z^2 - 2xy + 4xz - x + y = 11$$

at the point $(2,3,1)$.

The y coordinate where the tangent plane intersects the $y-$axis is

a) 1   b) 2   c) 3   d) 4   e) 5   f) 6   g) 7   h) 8

2.10) Find the equation of the tangent plane to the surface

$$x^2 + y^2 + z^2 - 2xy + 4xz - x + y = 11$$

at the point $(2,3,1)$.

The y coordinate where the tangent plane intersects the $y-$axis is

a) 1   b) 2   c) 3   d) 4   e) 5   f) 6   g) 7   h) 8

z defined implicitly $\Rightarrow$ Tangent Plane : simplify to $F(x,y,z)= C$

$$F_x(x_0,y_0,z_0)(x-x_0) + F_y(x_0,y_0,z_0)(y-y_0) + F_z(x_0,y_0,z_0)(z-z_0) = 0$$

$x^2 + y^2 + z^2 - 2xy + 4xz - x + y = 11$ @ the point $(2,3,1)$

The left side of the equation is $F(x,y,z)$

$F_x = 2x - 2y + 4z - 1$ $F_x(2,3,1) = 4-6+4-1 = 1$

$F_y = 2y - 2x + 1$ $F_y(2,3,1) = 6-4+1 = 3$

$F_z = 2z + 4x$ $F_z(2,3,1) = 2+8 = 10$

$1(x-2) + 3(y-3) + 10(z-1) = 0$

$x-2 + 3y - 9 + 10z - 10 = 0$

$\underline{x + 3y + 10z = 21}$ intersects   The equation

Tangent plane equation   y axis when   becomes $\boxed{y = 7}$

$x = z = 0$ $3y = 21$

2.11) Find the linearization of $f(x,y) = x^2 - xy + \dfrac{y^2}{2} + 3$

at the point $(3,2)$.

a) $x - y - z + 2 = 0$

e) $-3x - 3y - z - 5 = 0$

b) $4x - y - z + -2 = 0$

f) $2x - 7y - z = 0$

c) $2x + 2y - z - 3 = 0$

g) $x + 2y + 3z + 4 = 0$

d) $2x - y - z + 3 = 0$

h) $4x + 3y - 2z + 1 = 0$

2.11) Find the linearization of $f(x,y) = x^2 - xy + \dfrac{y^2}{2} + 3$

at the point $(3,2)$.

a) $x - y - z + 2 = 0$ e) $-3x - 3y - z - 5 = 0$

b) $4x - y - z + -2 = 0$ f) $2x - 7y - z = 0$

c) $2x + 2y - z - 3 = 0$ g) $x + 2y + 3z + 4 = 0$

d) $2x - y - z + 3 = 0$ h) $4x + 3y - 2z + 1 = 0$

$f(x,y) = x^2 - xy + \dfrac{y^2}{2} + 3$ at $(3,2)$

$f(3,2) = 9 - 6 + \dfrac{4}{2} + 3$
$f(3,2) = 3 + 2 + 3 = 8$

Linearization ⟹ Tangent Plane

when $z = f(x,y)$ explicitly we have
   the tangent plane equation as

$f_x = 2x - y$
$f_x(3,2) = 6 - 2 = 4$

$f_y = -x + y$
$f_y(3,2) = -3 + 2 = -1$

$z - f(x_0,y_0) = f_x(x_0,y_0)(x-x_0) + f_y(x_0,y_0)(y-y_0)$

$z - 8 = 4(x-3) + {}^-1(y-2)$

$z - 8 = 4x - 12 - y + 2$

$\boxed{0 = 4x - y - z - 2}$

2.12) Find the number of critical points that the function
$$f(x,y) = 2xe^x \sin y$$
has if $0 \le y \le 2\pi$.

a) 0   b) 1   c) 2   d) 3   e) 4   f) 5   g) 6   h) 7

2.12) Find the number of critical points that the function
$$f(x,y) = 2xe^x \sin y$$
has if $0 \le y \le 2\pi$.

a) 0   b) 1   c) 2   d) 3   e) 4   f) 5   g) 6   h) 7

$f(x,y) = 2xe^x \sin y$ Find how many critical points f has.

Critical points $\Rightarrow f_x = 0$ and $f_y = 0$ simultaneously

$f_x = \sin y \left[2e^x + 2xe^x \right] = \sin y \, 2e^x \left[1 + x \right]$

$f_y = 2xe^x \cdot \cos y$

$f_x = 0 \Rightarrow$ either $\sin y = 0 \Rightarrow y = 0, \pi, 2\pi$ since $0 \le y \le 2\pi$

$2e^x = 0 \Rightarrow e^x = 0$, this never happens

or $1 + x = 0 \Rightarrow x = -1$

and at the same time

$f_y = 0 \Rightarrow$ either $2x = 0 \Rightarrow x = 0$

$e^x = 0 \Rightarrow$ Never

or $\cos y = 0 \Rightarrow y = \frac{\pi}{2}, \frac{3\pi}{2}$

critical points

$\left(-1, \frac{\pi}{2} \right) \left(-1, \frac{3\pi}{2} \right)$

$(0,0) \; (0,\pi) \; (0,2\pi)$

Total number of critical points $\Big\}$ $\boxed{5}$

2.13) Find and classify all critical points for the function

$$f(x, y) = 27x^3 - y^3 - 18xy + 7$$

2.13) Find and classify all critical points for the function
$$f(x,y) = 27x^3 - y^3 - 18xy + 7$$

$f(x,y) = 27x^3 - y^3 - 18xy + 7$

Critical points: $f_x = 0$ AND $f_y = 0$

  or f_x and/or f_y DNE

$f_x = 81x^2 - 18y = 0$ $y = \dfrac{81x^2}{18} = \dfrac{9x^2}{2}$ $\left(\dfrac{9x^2}{2}\right)^2 = -6x$

$f_y = -3y^2 - 18x = 0$ $y^2 = \dfrac{-18x}{3} = -6x$ $\dfrac{81x^4}{4} = -6x$

$(0,0)$ and $\left(-\dfrac{2}{3}, 2\right)$ are the critical pts.

$f_{xx} = 162x$
$f_{yy} = -6y$ $D = (f_{xx})(f_{yy}) - (f_{xy})^2$
$f_{xy} = -18$ $D = (162x)(6y) - (-18)^2$

$D_{(0,0)} = -324 < 0$ $\boxed{\text{SADDLE POINT } (0,0,7)}$

$D_{\left(-\frac{2}{3},2\right)} = 162\left(-\frac{2}{3}\right)(-12) - 324$
$= 1296 - 324 > 0$ $\boxed{\text{LOCAL MAXIMUM}}$
$f_{xx} = -54 < 0$
$\boxed{\left(-\frac{2}{3}, 2, 15\right)}$

$\dfrac{81}{4}x^4 + 6x = 0$

$3x\left(\dfrac{27}{4}x^3 + 2\right) = 0$

$3x = 0$ $\dfrac{27}{4}x^3 = -2$

$x = 0$ $x^3 = -\dfrac{8}{27}$

$\Downarrow$ $x = -\dfrac{2}{3}$

$y = \dfrac{9}{2} \cdot 0$

$y = 0$ $\Downarrow$

$f(0,0) = 7$ $y = \dfrac{9}{2}\left(\dfrac{2}{3}\right)^2$

$(0,0,7)$ $y = \dfrac{9}{2} \cdot \dfrac{4}{9}$

$f\left(-\frac{2}{3},2\right) = 27 \cdot -\frac{8}{27} \cdot 8 - 18\left(-\frac{2}{3}\right) \cdot 2 + 7$ $y = 2$

$= -8 - 8 + 24 + 7$ $\left(-\frac{2}{3}, 2\right)$

$= 15$

$\left(-\frac{2}{3}, 2, 15\right)$

2.14) Let $f(x,y) = x^3 + y^3 + 3x^2 - 3y^2 - 8$.

Let a = the local minimum value of f and

b = the local maximum value of f. Find $a + b$.

a) 12   b) 16   c) 0   d) 4   e) 2   f) −4   g) −12   h) −16

2.14) Let $f(x,y) = x^3 + y^3 + 3x^2 - 3y^2 - 8$.

Let a = the local minimum value of f and

b = the local maximum value of f. Find $a+b$.

a) 12   b) 16   c) 0   d) 4   e) 2   f) −4   g) −12   h) −16

$f(x,y) = x^3 + y^3 + 3x^2 - 3y^2 - 8$

$f_x = 3x^2 + 6x \overset{set}{=} 0 \Rightarrow 3x(x+2) = 0 \begin{cases} x = 0 \\ x = -2 \end{cases}$

$f_y = 3y^2 - 6y \overset{set}{=} 0 \Rightarrow 3y(y-2) = 0 \begin{cases} y = 0 \\ y = 2 \end{cases}$

$(0,0) \quad (0,2)$
$(-2,0) \quad (-2,2)$

$f_{xx} = 6x + 6$
$f_{yy} = 6y - 6$
$f_{xy} = 0$

$D = (f_{xx})(f_{yy}) - (f_{xy})^2$

$D(0,0) = (6)(-6) = -36 < 0$ SADDLE POINT

$D(0,2) = (6)(6) = 36 > 0$ LOCAL MINIMUM
$f_{xx}(0,2) = 6 > 0$

$D(-2,0) = (-6)(-6) = 36 > 0$ LOCAL MAXIMUM
$f_{xx}(-2,0) = -6 < 0$

$D(-2,2) = (-6)(6) = -36 < 0$ SADDLE POINT

$f(0,2) = 0 + 8 + 0 - 12 - 8$
$f(0,2) = -12 \leftarrow$ the local minimum value

$a = -12$

$f(-2,0) = -8 + 0 + 12 - 0 - 8$
$f(-2,0) = -4 \leftarrow$ the local maximum value

$b = -4$

$a + b = -12 + -4$

$\boxed{a + b = -16}$

2.15) Find the point on the sphere

$$x^2 + y^2 + z^2 = 36$$

closest to $(1,2,2)$ and give the distance.

2.15) Find the point on the sphere

$$x^2 + y^2 + z^2 = 36$$

closest to $(1,2,2)$ and give the distance.

Minimize the distance a point (x,y,z) is to $(1,2,2)$

$d = \sqrt{(x-1)^2 + (y-2)^2 + (z-2)^2}$. To minimize this, we can minimize what's under the $\sqrt{\ }$.

Minimize

$f = (x-1)^2 + (y-2)^2 + (z-2)^2$ subject to the constraint

$g = x^2 + y^2 + z^2 - 36$

$F = f - \lambda g = (x-1)^2 + (y-2)^2 + (z-2)^2 - \lambda(x^2 + y^2 + z^2 - 36)$

$F_x = 2(x-1) - 2x\lambda = 0 \Rightarrow 2(x-1) = 2x\lambda \quad \lambda = \frac{2(x-1)}{2x} = \frac{x-1}{x}$

$F_y = 2(y-2) - 2y\lambda = 0 \Rightarrow 2(y-2) = 2y\lambda \quad \lambda = \frac{2(y-2)}{2y} = \frac{y-2}{y}$

$F_z = 2(z-2) - 2z\lambda = 0 \Rightarrow 2(z-2) = 2z\lambda \quad \lambda = \frac{2(z-2)}{2z} = \frac{z-2}{z}$

$\frac{x-1}{x} = \frac{y-2}{y} \Rightarrow xy - y = xy - 2x \Rightarrow y = 2x$

$\frac{y-2}{y} = \frac{z-2}{z} \Rightarrow yz - 2z = yz - 2y \Rightarrow y = z$ so $z = 2x$

$x^2 + y^2 + z^2 = 36$

$x^2 + (2x)^2 + (2x)^2 = 36$

$x^2 + 4x^2 + 4x^2 = 36$

$9x^2 = 36 \quad x^2 = 4$

$x = \pm 2$

$x = 2 \qquad x = -2$

$y = 4 \qquad y = -4$

$z = 4 \qquad z = -4$

$f(2,4,4) = 1 + 4 + 4 = 9$

$f(-2,-4,-4) = 9 + 36 + 36 = 81$

$d = \sqrt{9} @ (2,4,4)$
$d = 3 \leftarrow$ Abs Minimum

$d = \sqrt{81} @ (-2,-4,-4)$
$d = 9 \leftarrow$ Abs Maximum

2.16) Consider $f(x,y,z) = xy + xz$ subject to the constraint $x^2 + y^2 + z^2 = 32$.

Let A = the absolute minimum value of f and

B = the absolute maximum value of f.

Find $\dfrac{A}{B}$.

a) -4   e) 2

b) -2   f) 4

c) -1   g) 6

d) -6   h) 8

2.16) Consider $f(x,y,z) = xy + xz$ subject to the constraint $x^2 + y^2 + z^2 = 32$.

Let A = the absolute minimum value of f and

B = the absolute maximum value of f.

Find $\dfrac{A}{B}$.

a) -4   e) 2
b) -2   f) 4
c) -1   g) 6
d) -6   h) 8

$f(x,y) = xy + xz$
subject to
$g(x,y) = x^2 + y^2 + z^2 - 32$

$F = f - \lambda g = xy + xz - \lambda(x^2 + y^2 + z^2 - 32)$

$F_x = y + z - 2x\lambda = 0 \qquad y + z = 2x\lambda$

$F_y = x - 2y\lambda = 0 \qquad\qquad x = 2y\lambda$

$F_z = x - 2z\lambda = 0 \qquad\qquad x = 2z\lambda$

$A = 16\sqrt{2}$
$B = 16\sqrt{2}$

$\boxed{\dfrac{A}{B} = -1}$

$\dfrac{y+z}{2x} = \dfrac{x}{2y} = \dfrac{x}{2z}$

$2y^2 + 2yz = 2x^2 \qquad 2z x = 2y x$

$2y^2 + 2y^2 = 2x^2 \qquad y = z$

$4y^2 = 2x^2$

$2y^2 = x^2$

$x^2 + y^2 + z^2 = 32$

$2y^2 + y^2 + y^2 = 32$

$4y^2 = 32$

$y^2 = 8$

$y = \pm 2\sqrt{2}$

$y = 2\sqrt{2}$
$z = 2\sqrt{2}$
$x^2 = 2y^2$
$x^2 = 2(2\sqrt{2})^2$
$x^2 = 2 \cdot 8$
$x^2 = 16$
$x = 4 \quad x = -4$
$P(4, 2\sqrt{2}, 2\sqrt{2})$
$Q(-4, 2\sqrt{2}, 2\sqrt{2})$

$y = -2\sqrt{2}$
$z = -2\sqrt{2}$
$x^2 = 2y^2$
$x^2 = d \cdot 8$
$x^2 = 16$
$x = 4 \quad x = -4$
$R(4, -2\sqrt{2}, -2\sqrt{2})$
$S(-4, -2\sqrt{2}, -2\sqrt{2})$

P: $f = 8\sqrt{2} + 8\sqrt{2}$
$= 16\sqrt{2}$

Q: $f = -8\sqrt{2} - 8\sqrt{2}$
$= -16\sqrt{2}$

R: $f = -8\sqrt{2} - 8\sqrt{2}$
$= 16\sqrt{2}$

S: $f = 8\sqrt{2} + 8\sqrt{2}$
$= 16\sqrt{2}$

2.17) Minimize $f(r,h) = 2\pi r^2 + 2\pi rh$ subject to the constraint $r^2 h = 54$ and $r > 0$ and $h > 0$.

2.17) Minimize $f(r,h) = 2\pi r^2 + 2\pi rh$ subject to the constraint $r^2h = 54$ and $r > 0$ and $h > 0$.

Minimize
$$f(r,h) = 2\pi r^2 + 2\pi rh$$
Subject to
$$g(r,h) = r^2h - 54$$

$$F = f - \lambda g = 2\pi r^2 + 2\pi rh - \lambda(r^2h - 54)$$

$$F_r = 4\pi r + 2\pi h - 2rh\lambda = 0 \Rightarrow 4\pi r + 2\pi h = 2rh\lambda$$

$$F_h = 2\pi r - r^2\lambda = 0 \Rightarrow 2\pi r = r^2\lambda$$

$$\lambda = \frac{4\pi r + 2\pi h}{2rh} = \frac{2\pi r}{r^2}$$

$$4\pi r^3 + 2\pi r^2 h = 4\pi r^2 h$$

$$4\pi r^3 - 2\pi r^2 h = 0$$

$$2\pi r^2(2r - h) = 0$$

$\underset{r = 0}{\downarrow}$ $h = 2r$

Not on g

$$r^2 h = 54$$

$$r^2(2r) = 54$$

$$2r^3 = 54$$

$$r^3 = 27$$

$\rightarrow r = 3$
$h = 2 \cdot 3$
$h = 6$

$$f(3,6) = 2\pi \cdot 9 + 2\pi \cdot 18$$
$$= 18\pi + 36\pi$$
$$= 54\pi, \text{ this is an extreme value.}$$

$\underset{abs. max}{\nearrow}$ or $\underset{abs. min.}{\searrow}$

$r > 0$ Find another combination that
$h > 0$ satisfies $r^2 h = 54$ ex. $\underset{h=1}{r=54}$
Plug this into f

$$f(54,1) > f(3,6)$$

so $\boxed{f(3,6) = 54\pi}$ must be the absolute minimum

2.18) Find the volume of the solid lying under the plane $z = 4 - 2x - y$

and above the rectangle $R = \{(x, y) | 0 \le x \le 1,\ \text{and}\ 0 \le y \le 2\}$.

a) 0 b) 4 c) 8 d) 12
e) 2 f) 6 g) 10 h) 16

2.18) Find the volume of the solid lying under the plane $z = 4 - 2x - y$ and above the rectangle $R = \{(x, y) | 0 \le x \le 1, \text{ and } 0 \le y \le 2\}$.

a) 0

b) 4

c) 8

d) 12

e) 2

f) 6

g) 10

h) 16

Volume under
$$z = 4 - 2x - y$$
above $R: 0 \le x \le 1$
$$0 \le y \le 2$$

$$V = \iint_R (4 - 2x - y) \, dA = \int_0^2 \int_0^1 (4 - 2x - y) \, dx \, dy$$

Take anti-deriv.
with respect to x
(hold y constant)

$$= \int_0^2 (4x - x^2 - yx)\Big|_{x=0}^{x=1} \, dy = \int_0^2 \Big[(4 - 1 - y) - 0\Big] \, dy$$

$$= \int_0^2 (3 - y) \, dy = \left(3y - \frac{y^2}{2}\right)\Big|_0^2 = \left(6 - \frac{4}{2}\right) - 0 = 6 - 2 = \boxed{4}$$

2.19) Evaluate the integral

$$\int_0^3 \int_{\sqrt{\frac{x}{3}}}^1 e^{y^3}\,dy\,dx$$

a) e

b) $\dfrac{1}{e}$

c) $3e$

d) $\dfrac{3}{e}$

e) $e-1$

f) $1-e^3$

2.19) Evaluate the integral

$$\int_0^3 \int_{\sqrt{\frac{x}{3}}}^1 e^{y^3}\, dy\, dx$$

a) e

b) $\dfrac{1}{e}$

c) $3e$

d) $\dfrac{3}{e}$

e) $e-1$

f) $1-e^3$

$$\int_0^3 \int_{\sqrt{\frac{x}{3}}}^1 \underbrace{e^{y^3}\, dy\, dx}$$

can't integrate in y first ⟹ Switch order of integration

lower bound $\sqrt{\frac{x}{3}} \le y \le 1$ upper bound $dy\,dx \quad dx\,dy$

$0 \le x \le 3$

$y = \sqrt{\frac{x}{3}}$

$y^2 = \frac{x}{3}$

$x = 3y^2$

$$= \int_0^1 \int_0^{3y^2} e^{y^3}\, dx\, dy$$

$$= \int_0^1 e^{y^3}[x]_{x=0}^{x=3y^2}\, dy$$

$$= \int_0^1 e^{y^3}(3y^2 - 0)\, dy$$

$$= \int_0^1 3y^2 e^{y^3}\, dy \qquad u = y^3$$
$$du = 3y^2\, dy$$
$$\int e^u du = e^u$$

$$= e^{y^3}\Big|_0^1$$

$$= e^1 \cdot e^0 = \boxed{e - 1}$$

2.20) Evaluate

$$\int_{0}^{\pi^{1/3}}\int_{y^2}^{\pi^{2/3}} 3\sin\left(x\sqrt{x}\right)dxdy$$

a) 0   b) 1   c) 2   d) 3   e) 4   f) $\dfrac{8}{3}$

2.20) Evaluate

$$\int_{0}^{\pi^{1/3}} \int_{y^2}^{\pi^{2/3}} 3\sin\left(x\sqrt{x}\right) dx\,dy$$

a) 0   b) 1   c) 2   d) 3   e) 4   f) $\dfrac{8}{3}$

You can't integrate w.r.t. x first ⇒ switch the order
of integration to
$dx\,dy$ → $dy\,dx$

dx dy

$x = y^2$

LB

UB

$\pi^{1/3}$

$\pi^{2/3}$

dy dx

UB

$y = \sqrt{x}$

LB

$\pi^{2/3}$

$^{LB}y^2 \le x \le \pi^{2/3}{}^{UB}$

$0 \le y \le \pi^{1/3}$

$^{LB}0 \le y \le \sqrt{x}^{UB}$

$0 \le x \le \pi^{2/3}$

$$\int_{0}^{\pi^{2/3}} \int_{0}^{\sqrt{x}} 3\sin(x\sqrt{x})\,dy\,dx$$
\underbrace{\phantom{3\sin(x\sqrt{x})}}_{\text{constant w.r.t. }y}

$$\int_{0}^{\pi^{2/3}} 3\sin(x\sqrt{x})\left[y\right]_{0}^{\sqrt{x}} dx$$
$(\sqrt{x}-0)$

$$\int_{0}^{\pi^{2/3}} 3\sqrt{x}\,\sin(x^{3/2})\,dx$$

$$\int_{0}^{\pi^{2/3}} 3\sqrt{x}\,\sin(x^{3/2})\,dx$$

$u = x^{3/2}$

$du = \frac{3}{2}x^{1/2}\,dy$

$3\left(\frac{2}{3}du = x^{1/2}dx\right)$

$2du = 3x^{1/2}dx$

$2\int \sin u \, du = -2\cos u$

$$= -2\left[\cos(x^{3/2})\right]_{0}^{\pi^{2/3}}$$

$$= -2\left[\underset{-1}{\cos \pi} - \underset{1}{\cos 0}\right] = -2[-2] = \boxed{4}$$

2.21) Integrate

$$\iint\limits_{R} (6x + 9y)\, dA$$

where R is the region in the first quadrant bounded by the circles $x^2 + y^2 = 1$ and $x^2 + y^2 = 4$.

A) 1   B) 12   C) 20   D) 32

E) 35   F) 42   G) 48   H) None of these

2.21) Integrate

$$\iint\limits_{R} (6x + 9y)\, dA$$

where R is the region in the first quadrant bounded by the circles $x^2 + y^2 = 1$ and $x^2 + y^2 = 4$.

A) 1   B) 12   C) 20   D) 32
E) 35   F) 42   G) 48   H) None of these

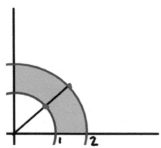

$$\int_{0}^{\pi/2}\int_{1}^{2}\left[6(r\cos\theta) + 9(r\sin\theta)\right] r\, dr\, d\theta$$

$$= \int_{0}^{\pi/2}\int_{1}^{2}\left(6\cos\theta + 9\sin\theta\right) r^2\, dr\, d\theta$$

Circular region ⟹ do the integral
in polar

$1 \le r \le 2 \qquad x = r\cos\theta$
$0 \le \theta \le \dfrac{\pi}{2} \qquad y = r\sin\theta$
$\qquad\qquad dA = r\, dr\, d\theta$

$$= \int_{0}^{\pi/2}(6\cos\theta + 9\sin\theta)\left[\frac{r^3}{3}\right]_{1}^{2} d\theta$$

$$= \int_{0}^{\pi/2}(6\cos\theta + 9\sin\theta)\left(\frac{8}{3} - \frac{1}{3}\right) d\theta$$

$$= \int_{0}^{\pi/2}\frac{7}{3}\cdot 3\left[2\cos\theta + 3\sin\theta\right] d\theta$$

$$= 7\left[2\sin\theta - 3\cos\theta\right]_{0}^{\frac{\pi}{2}} = 7\left[(2 - 0) - (0 - 3)\right] = \boxed{35}$$

2.22) Calculate the volume under the surface $f(x, y) = \dfrac{2\sin y}{y}$ and

above the region R in the xy – plane bounded by the lines

$y = x,\ y = \dfrac{1}{2}x,\ y = \dfrac{\pi}{3},$ and $y = \dfrac{\pi}{2}.$

a) 0   b) 1   c) 2   d) 3   e) 4   f) 5

2.22) Calculate the volume under the surface $f(x,y) = \dfrac{2\sin y}{y}$ and

above the region R in the xy – plane bounded by the lines

$y = x$, $y = \dfrac{1}{2}x$, $y = \dfrac{\pi}{3}$, and $y = \dfrac{\pi}{2}$.

a) 0   b) 1   c) 2   d) 3   e) 4   f) 5

The integrand $\dfrac{2\sin y}{y}$ cannot be integrated w.r.t. y

$\Rightarrow$ don't do $dy\,dx$
The region is also dictating not to do it in $dy\,dx$

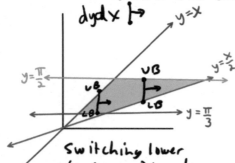

$dy\,dx \mapsto$

$dx\,dy$

Switching lower
(and upper) bounds
avoid this if possible

$y \le x \le 2y$
$\dfrac{\pi}{3} \le y \le \dfrac{\pi}{2}$

$$\int_{\frac{\pi}{3}}^{\frac{\pi}{2}}\int_{y}^{2y}\frac{2\sin y}{y}\,dx\,dy \;=\; \int_{\frac{\pi}{3}}^{\pi/2}\frac{2\sin y}{y}\,[x]_{y}^{2y}\,dy \;=\; \int_{\frac{\pi}{3}}^{\pi/2}\frac{2\sin y}{y}\cdot y\,dy$$

$$\qquad\qquad\qquad\qquad (2y-y)$$

$$=\; 2\int_{\frac{\pi}{3}}^{\frac{\pi}{2}}\sin y\,dy \;=\; 2\,[-\cos y]_{\frac{\pi}{3}}^{\pi/2} \;=\; 2\left[\cos\frac{\pi}{2} - -\cos\frac{\pi}{3}\right] \;=\; \boxed{1}$$

$$\qquad\qquad\qquad\qquad\qquad\qquad 0 \qquad \tfrac{1}{2}$$

$$\qquad\qquad\qquad\qquad\qquad 2\,(1/2)$$

2.23) Find the volume between the planes

$x + y + z = 4$ and $4x + y + 4z = 16$

in the first octant.

a) 0   b) 4   c) 8   d) 16   e) 24   f) 32   g) 42   h) 64

2.23) Find the volume between the planes

$x + y + z = 4$ and $4x + y + 4z = 16$

in the first octant.

a) 0   b) 4   c) 8   d) 16   e) 24   f) 32   g) 42   h) 64

$x+y+z=4$ $4x+y+4z=16$
$x=y=0 \Rightarrow z=4$ $x=y=0 \Rightarrow z=4$
$x=z=0 \Rightarrow y=4$ $x=z=0 \Rightarrow y=16$
$y=z=0 \Rightarrow x=4$ $y=z=0 \Rightarrow x=4$
Plane A   Plane B

* The lower z changes
  - for part of the time the lower z is plane A and the rest of the time the lower z is the xy plane.

Integrate in y first

* Similarly the lower x changes

$$\int_0^4 \int_0^{4-x} \int_{4-x-z}^{4(4-x-z)} 1 \, dy \, dz \, dx$$

The Lower y is always Plane A   Solve for y $y = 4-x-z$
The Upper y is always Plane B $y = 16-4x-4z$
$y = 4(4-x-z)$

The shadow region in the xz plane is:

$b=4, m=-1$
$z = -x+4$
$z = 4-x$
$dz \, dx$
(can also be done $dx \, dz$)

$$= \int_0^4 \int_0^{4-x} [z]_{4-x-z}^{4(4-x-z)} \, dz \, dx = \int_0^4 \int_0^{4-x} 3[4-x-z] \, dz \, dx$$

$$= 3 \int_0^4 \int_0^{4-x} (4-x-z) \, dz \, dx = 3 \int_0^4 \left[4z - xz - \frac{z^2}{2}\right]_0^{4-x} dx = 3 \int_0^4 \left(4(4-x) - x(4-x) - \frac{(4-x)^2}{2}\right) dx$$

$\underbrace{\qquad}_{(4-x)(4-x)}$

$(4-x)^2 - \frac{1}{2}(4-x)^2 = \frac{1}{2}(4-x)^2$

$$= 3 \int_0^4 \frac{1}{2}(16-8x+x^2) \, dx = \frac{3}{2}\left[16x - 4x^2 + \frac{x^3}{3}\right]_0^4 = \frac{3}{2}\left(64-64+\frac{64}{3}\right) = \frac{3}{2} \cdot \frac{64}{3} = \boxed{32}$$

2.24) Find the volume of the solid bounded l

$$z = 5 - 5\left(x^2 + y^2\right) \text{ and } z = \left(x^2 + y^2\right)^2 - 1.$$

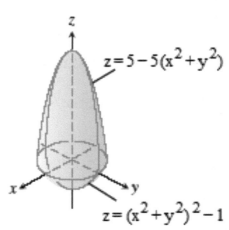

a) $\dfrac{13\pi}{3}$ b) $\dfrac{17\pi}{3}$ c) $\dfrac{19\pi}{6}$ d) $\dfrac{13\pi}{6}$

e) $\dfrac{17\pi}{6}$ f) $\dfrac{13\pi}{2}$ g) $\dfrac{19\pi}{2}$ h) None of the above

2.24) Find the volume of the solid bounded b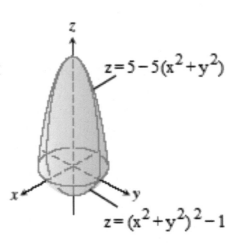

$z = 5 - 5(x^2 + y^2)$ and $z = (x^2 + y^2)^2 - 1$.

$z = 5 - 5(x^2 + y^2)$

a) $\dfrac{13\pi}{3}$ b) $\dfrac{17\pi}{3}$ c) $\dfrac{19\pi}{6}$ d) $\dfrac{13\pi}{6}$

e) $\dfrac{17\pi}{6}$ f) $\dfrac{13\pi}{2}$ g) $\dfrac{19\pi}{2}$ h) None of the above

$z = (x^2 + y^2)^2 - 1$

Definite upper and lower z $\Big\}$ Do the integral
together with circular in cylindrical
region in the xy plane

$z = 5 - 5(x^2 + y^2) \Rightarrow z = 5 - 5r^2$ oR $z = 5(1 - r^2)$

$z = (x^2 + y^2)^2 - 1 \Rightarrow z = (r^2)^2 - 1$ oR $z = r^4 - 1$

$\displaystyle\int_0^{2\pi}\int_0^1 \int_{r^4-1}^{5(1-r^2)} r\,dz\,dr\,d\theta$

xy plane: $z = 0$

$0 = r^4 - 1 \atop \text{oR} \atop 0 = 1 - r^2$ $\Big\} \Rightarrow r = 1$ unit circle

(the entire unit circle)

$0 \le r \le 1$

$0 \le \theta \le 2\pi$

$\displaystyle\int_0^{2\pi}\int_0^1 r\,[z]_{r^4-1}^{5(1-r^2)} \,dr\,d\theta$

$\displaystyle\int_0^{2\pi}\int_0^1 r\left[\underbrace{5(1-r^2)}_{5-5r^2} - \underbrace{(r^4-1)}_{6-5r^2-r^4}\right]^{5-5r^2-r^4+1} dr\,d\theta =$

$\displaystyle\int_0^{2\pi}\int_0^1 (6r - 5r^3 - r^5)\,dr\,d\theta = \int_0^{2\pi}\left(3r^2 - \tfrac{5r^4}{4} - \tfrac{r^6}{6}\right)\Big|_0^1 d\theta$

$= \displaystyle\int_0^{2\pi}\left(3 - \tfrac{5}{4} - \tfrac{1}{6}\right)d\theta = \tfrac{36-15-2}{12}\int_0^{2\pi}d\theta = \tfrac{19}{12}\cdot 2\pi$

$= \boxed{\dfrac{19\pi}{6}}$

2.25) The process involved in finding the volume of the solid bounded above by the sphere $\rho = 14$ and below by the plane $z = 7$ requires finding bounds on $\rho, \phi,$ and θ.

Set up **BUT DO NOT SOLVE** this integral.

$$\int_{}^{} \int_{}^{} \int_{}^{} \rho^2 \sin \phi \, d\rho \, d\phi \, d\theta$$

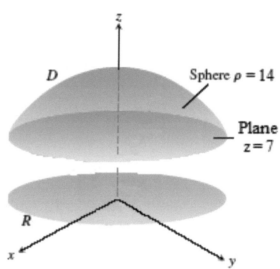

D

Sphere $\rho = 14$

Plane $z = 7$

R

2.25) The process involved in finding the volume of the solid bounded above by the sphere $\rho = 14$ and below by the plane $z = 7$ requires finding bounds on $\rho, \phi,$ and θ.

Set up **BUT DO NOT SOLVE** this integral.

$$\int_{\square}^{\square} \int_{\square}^{\square} \int_{\square}^{\square} \rho^2 \sin \phi \, d\rho \, d\phi \, d\theta$$

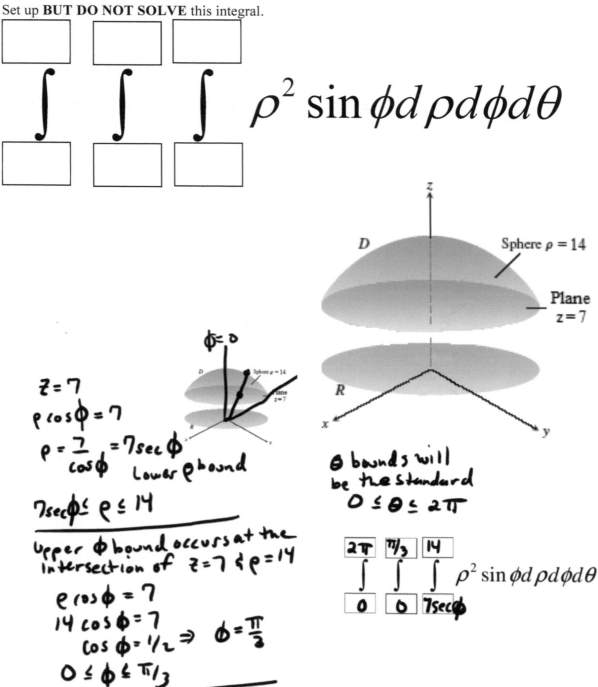

$\phi = 0$

$z = 7$

$\rho \cos \phi = 7$

$\rho = \dfrac{7}{\cos \phi} = 7 \sec \phi$ Lower ρ bound

$7 \sec \phi \leq \rho \leq 14$

Upper ϕ bound occurs at the intersection of $z = 7$ & $\rho = 14$

$\rho \cos \phi = 7$

$14 \cos \phi = 7$

$\cos \phi = 1/2 \implies \phi = \dfrac{\pi}{3}$

$0 \leq \phi \leq \pi/3$

θ bounds will be the standard

$0 \leq \theta \leq 2\pi$

Sphere $\rho = 14$

Plane $z = 7$

$$\int_{0}^{2\pi} \int_{0}^{\pi/3} \int_{7\sec\phi}^{14} \rho^2 \sin \phi \, d\rho \, d\phi \, d\theta$$

2.26) Find the volume of the solid pictured below:

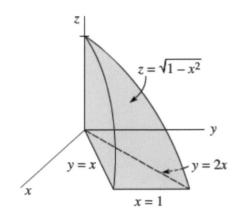

$z = \sqrt{1 - x^2}$

$y = x$

$y = 2x$

$x = 1$

a) 4   b) 3   c) 2   d) 1   e) $\dfrac{1}{3}$ f) $\dfrac{2}{5}$ g) $\dfrac{1}{2}$ h) $\dfrac{1}{4}$

2.26) Find the volume of the solid pictured below:

$z = \sqrt{1-x^2}$

y

$y = x$

$y = 2x$

$x = 1$

a) 4   b) 3   c) 2   d) 1   e) $\dfrac{1}{3}$ f) $\dfrac{2}{5}$ g) $\dfrac{1}{2}$ h) $\dfrac{1}{4}$

Definite upper and lower z
together with xy plane region
not being circular
in nature

$\left.\vphantom{\begin{array}{c}a\\b\\c\\d\end{array}}\right\}$ Do the Integral
in Cartesian

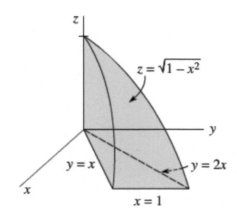

$y=2x$ $0 \le z \le \sqrt{1-x^2}$

$y=x$ $x \le y \le 2x$

$dydx$ $0 \le x \le 1$

$x=1$

$$V = \int_0^1 \int_x^{2x} \int_0^{\sqrt{1-x^2}} 1\, dz\, dy\, dx = \int_0^1 \int_x^{2x} \Big[z\Big]_0^{\sqrt{1-x^2}} dy\, dx$$

$$= \int_0^1 \int_x^{2x} \sqrt{1-x^2}\, dy\, dx = \int_0^1 \sqrt{1-x^2}\, \Big[y\Big]_x^{2x} dx$$

$$= \int_0^1 \sqrt{1-x^2}\,(2x-x)dx = \int_0^1 x\sqrt{1-x^2}\, dx$$

$u = 1-x^2$

$du = -2x\, dx$

$-\tfrac{1}{2}du = x\, dx$

$= -\tfrac{1}{2}\int u^{1/2} du$

$= -\tfrac{1}{2}\cdot\tfrac{2}{3} u^{3/2}$

$$= -\tfrac{1}{3}(1-x^2)^{3/2}\Big|_0^1$$

$$= -\tfrac{1}{3}\Big[0 - 1^{3/2}\Big]$$

$$= \boxed{\dfrac{1}{3}}$$

2.27) Evaluate

$$\int_{0}^{2}\int_{0}^{\sqrt{4-x^2}}\int_{0}^{1}\left(x^2+y^2\right)^{3/2}\,dz\,dy\,dx$$

a) $\dfrac{7\pi}{2}$ b) 3π c) $\dfrac{15\pi}{4}$ d) $\dfrac{12\pi}{5}$ e) $\dfrac{13\pi}{3}$ f) $\dfrac{14\pi}{5}$ g) 4π h) $\dfrac{16\pi}{5}$

2.27) Evaluate

$$\int_0^2 \int_0^{\sqrt{4-x^2}} \int_0^1 \left(x^2+y^2\right)^{3/2} \, dz\,dy\,dx$$

a) $\dfrac{7\pi}{2}$ b) 3π c) $\dfrac{15\pi}{4}$ d) $\dfrac{12\pi}{5}$ e) $\dfrac{13\pi}{3}$ f) $\dfrac{14\pi}{5}$ g) 4π h) $\dfrac{16\pi}{5}$

$x^2+y^2 \Rightarrow$ switch into cylindrical

$0 \le z \le 1$

$0 \le y \le \sqrt{4-x^2}$ $\}$ upper half $y=\sqrt{4-x^2} \Rightarrow x^2+y^2=4$
Circle of radius 2

$0 \le x \le 2$ 1st quadrant

$0 \le r \le 2$
$0 \le \theta \le \pi/2$

$$V = \int_0^{\pi/2} \int_0^2 \int_0^1 \left(r^2\right)^{3/2} \, dz\, r\, dr\, d\theta = \int_0^{\pi/2} \int_0^2 r^4 \left(z\right)_0^1 dr\, d\theta$$

$\underbrace{r^3}$

$$= \int_0^{\pi/2} \int_0^2 r^4 \, dr\, d\theta = \int_0^{\pi/2} \left(\frac{r^5}{5}\right)_0^2 d\theta = \frac{32}{5} \int_0^{\pi/2} d\theta$$

$$= \frac{32}{5} \cdot \frac{\pi}{2} = \boxed{\frac{16\pi}{5}}$$

2.28) Evaluate

$$\int_{0}^{2\pi} \int_{\arctan(1/3)}^{\frac{\pi}{4}} \int_{0}^{3\sec\phi} \rho^2 \sin\phi \, d\rho d\phi d\theta$$

a) 0   b) π c) 2π d) 3π e) 4π f) 6π g) 8π h) 12π

2.28) Evaluate

$$\int_{0}^{2\pi} \int_{\arctan(1/3)}^{\frac{\pi}{4}} \int_{0}^{3\sec\phi} \rho^2 \sin\phi \, d\rho\,d\phi\,d\theta$$

a) 0   b) π c) 2π d) 3π e) 4π f) 6π g) 8π h) 12π

$$\int_{0}^{2\pi} \int_{\arctan(1/3)}^{\pi/4} \left[\frac{\rho^3}{3}\right]_{0}^{3\sec\phi} \sin\phi \, d\phi\,d\theta = \int_{0}^{2\pi} \int_{\arctan(1/3)}^{\pi/4} 9 \sec^3\phi \sin\phi \, d\phi\,d\theta$$

$$\underbrace{\sec^3\phi \cdot \frac{1}{\cos\phi} \cdot \sin\phi}$$
$$\sec^2\phi \tan\phi$$

$$= 9 \int_{0}^{2\pi} \int_{\arctan(1/3)}^{\pi/4} \tan\phi \sec^2\phi \, d\phi\,d\theta$$

$u = \tan\phi$
$du = \sec^2\phi \, d\phi$
$\int u \, du = \frac{u^2}{2}$

$\tan(\arctan 1/3) = \frac{1}{3}$

$$= 9 \int_{0}^{2\pi} \frac{1}{2}(\tan\phi)^2 \Big|_{\arctan\frac{1}{3}}^{\pi/4} d\theta$$

$$= 9 \cdot \frac{1}{2} \cdot \left(1 - \left(\tfrac{1}{3}\right)^2\right) \cdot \int_{0}^{2\pi} d\theta$$

$$= \frac{9}{2} \cdot \frac{8}{9} \cdot 2\pi \quad = \boxed{8\pi}$$

2.29) Find the volume of the region in the first octant bounded by the coordinate planes, the plane $y = 1 - x$,

and the surface $z = \cos\left(\dfrac{\pi}{2}x\right)$, $0 \le x \le 1$.

a) π

b) $\dfrac{4}{\pi^2}$

c) $\dfrac{4}{\pi}$

d) $\dfrac{2}{\pi}$

e) $\dfrac{8}{\pi^3}$

f) 1

2.29) Find the volume of the region in the first octant bounded by the coordinate planes, the plane $y = 1 - x$, and the surface $z = \cos\left(\dfrac{\pi}{2}x\right)$, $0 \le x \le 1$.

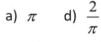

a) π d) $\dfrac{2}{\pi}$

b) $\dfrac{4}{\pi^2}$ e) $\dfrac{8}{\pi^3}$

c) $\dfrac{4}{\pi}$ f) 1

$0 \le z \le \cos\left(\frac{\pi}{2}x\right)$
$0 \le y \le 1-x$
$0 \le x \le 1$

$y = 1-x$

$$\int_0^1 \int_0^{1-x} \int_0^{\cos\left(\frac{\pi}{2}x\right)} 1 \; dz \, dy \, dx = \int_0^1 \int_0^{1-x} [z]_0^{\cos\left(\frac{\pi}{2}x\right)} dy\, dx$$

$$= \int_0^1 \int_0^{1-x} \cos\left(\tfrac{\pi}{2}x\right) dy\, dx = \int_0^1 \cos\tfrac{\pi}{2}x \cdot [y]_0^{1-x} dx$$

$$= \int_0^1 (1-x)\cos\left(\tfrac{\pi}{2}x\right) dx$$

$u = 1-x \qquad dv = \cos\tfrac{\pi}{2}x$
$du = -dx \qquad v = \tfrac{2}{\pi}\sin\left(\tfrac{\pi}{2}x\right)$

$uv - \int v\,du$

$$= \tfrac{2}{\pi}(1-x)\sin\left(\tfrac{\pi}{2}x\right) + \int \tfrac{2}{\pi}\sin\left(\tfrac{\pi}{2}x\right)dx$$

$$= \tfrac{2}{\pi}(1-x)\sin\left(\tfrac{\pi}{2}x\right) - \tfrac{4}{\pi^2}\cos\left(\tfrac{\pi}{2}x\right) \Big|_0^1$$

$$= \left(0 - \tfrac{4}{\pi^2}\cos\tfrac{\pi}{2}\right) - \left(\tfrac{2}{\pi}\sin 0 - \tfrac{4}{\pi^2}\cos 0\right) = \boxed{\dfrac{4}{\pi^2}}$$

2.30) Find the volume of the solid graphed below.

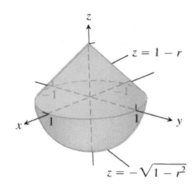

a) π d) 2π

b) 1 e) 2

c) $\dfrac{\pi}{2}$ f) $\dfrac{\pi}{4}$

2.30) Find the volume of the solid graphed below.

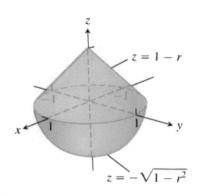

$z = 1 - r$

$z = -\sqrt{1 - r^2}$

a) π d) 2π

b) 1 e) 2

c) $\dfrac{\pi}{2}$ f) $\dfrac{\pi}{4}$

Defined upper and lower z ¿ — Do the integral
circular region in xy plane — in cylindrical

$$\int_0^{2\pi} \int_0^1 \int_{-\sqrt{1-r^2}}^{1-r} r \, dz \, dr \, d\theta = \int_0^{2\pi} \int_0^1 r \left[z\right]_{-\sqrt{1-r^2}}^{1-r} dr \, d\theta$$

$$= \int_0^{2\pi} \int_0^1 r\left((1-r) - -\sqrt{1-r^2}\right) dr \, d\theta = \int_0^{2\pi} \int_0^1 r - r^2 + r\sqrt{1-r^2} \, dr \, d\theta$$

$u = 1 - r^2$
$du = -2r \, dr$
$-\frac{1}{2} du = r \, dr$
$\quad \frac{-1}{2} \int u^{1/2} du$
$\quad \frac{-1}{2} \cdot \frac{2}{3} u^{3/2}$
$\quad \frac{-1}{3} u^{3/2}$

$$= \int_0^{2\pi} \left[\frac{r^2}{2} - \frac{r^3}{3} + \frac{-1}{3}(1-r^2)^{3/2} \right]_0^1 d\theta$$

$$= \int_0^{2\pi} \left(\left(\tfrac{1}{2} - \tfrac{1}{3} - \tfrac{1}{3}(0)\right) - \left(0 - 0 - \tfrac{1}{3}(1)\right)\right) d\theta = \tfrac{1}{2}\int_0^{2\pi} d\theta = \boxed{\pi}$$

2.31) Evaluate the integral

$$\int_{-2}^{2} \int_{-\sqrt{4-x^2}}^{\sqrt{4-x^2}} \int_{0}^{\sqrt{4-x^2-y^2}} e^{-\left(x^2+y^2+z^2\right)^{3/2}} \, dz\,dy\,dx$$

a) e

b) $\dfrac{2\pi}{3}\left(1-e^{-8}\right)$

c) $\dfrac{\pi}{3}\left(1-e\right)$

d) $\dfrac{\pi}{6}\left(1-e^2\right)$

e) $\dfrac{2\pi}{9}\left(1-e^{-4}\right)$

f) 1

2.31) Evaluate the integral

$$\int_{-2}^{2}\int_{-\sqrt{4-x^2}}^{\sqrt{4-x^2}}\int_{0}^{\sqrt{4-x^2-y^2}} e^{-\left(x^2+y^2+z^2\right)^{3/2}} \, dz\,dy\,dx$$

a) e

b) $\dfrac{2\pi}{3}\left(1-e^{-8}\right)$

c) $\dfrac{\pi}{3}\left(1-e\right)$

d) $\dfrac{\pi}{6}\left(1-e^2\right)$

e) $\dfrac{2\pi}{9}\left(1-e^{-4}\right)$

f) 1

$x^2+y^2+z^2 \Rightarrow$ Do the integral in spherical

$0 \le z \le \sqrt{4+x^2-y^2}$

lowest z is the xy plane

$-\sqrt{4-x^2} \le y \le \sqrt{4-x^2}$

$y = \pm\sqrt{4-x^2}$

$x^2+y^2 = 4$ circle of radius 2

$-2 \le x \le 2 \Rightarrow$ all of the circle

$0 \le \rho \le 2$
$0 \le \phi \le \frac{\pi}{2}$
$0 \le \theta \le 2\pi$

$z = \sqrt{4-x^2-y^2}$

$x^2+y^2+z^2 = 4$
sphere of radius 2
$\rho = 2$

$\phi = 0$

$\phi = \frac{\pi}{2}$

$\displaystyle\int_0^{2\pi}\int_0^{\pi/2}\int_0^{2} e^{-(\rho^2)^{3/2}} \cdot \rho^2 \sin\phi \, d\rho\, d\phi\, d\theta$

$\displaystyle\int_0^{2\pi}\int_0^{\pi/2}\int_0^{2} \rho^2 e^{-\rho^3} \, d\rho\, d\phi\, d\theta$

$u = -\rho^3$
$du = -3\rho^2\, d\rho$
$-\frac{1}{3} du = \rho^2 d\rho$

$-\frac{1}{3}\int e^u du$
$= -\frac{1}{3}e^u$

$\displaystyle\int_0^{2\pi}\int_0^{\pi/2} -\frac{1}{3}e^{-\rho^3}\Big|_0^{2} \sin\phi\, d\phi\, d\theta = \frac{-1}{3}(e^{-8}-1)\int_0^{2\pi}\int_0^{\pi/2}\sin\phi\, d\phi\, d\theta$

$= -\frac{1}{3}(e^{-8}-1)\left(-\cos\phi\right]_0^{\pi/2} \cdot \int_0^{2\pi} d\theta = \frac{-1}{3}(e^{-8}-1)\cdot 2\pi = \boxed{\dfrac{2\pi}{3}\left(1-e^{-8}\right)}$

$(0--1)$

$\underbrace{}_{2\pi}$

2.32) Evaluate

$$\iint_R \frac{4y^2}{x} \, dA$$

where R is the region bounded by the graphs

$$y = x^2, \quad y = \frac{1}{2}x^2, \quad x = y^2, \quad \text{and} \quad x = \frac{1}{2}y^2.$$

Let $u = \frac{x^2}{y}$ and $v = \frac{y^2}{x}$.

a) 6   b) 5   c) 4   d) 3   e) 2   f) 1   g) $\dfrac{1}{2}$ h) $\dfrac{1}{4}$

2.32) Evaluate

$$\iint_R \frac{4y^2}{x} \, dA$$

where R is the region bounded by the graphs

$$y = x^2, \ y = \frac{1}{2}x^2, \ x = y^2, \text{ and } x = \frac{1}{2}y^2.$$

Let $u = \frac{x^2}{y}$ and $v = \frac{y^2}{x}$.

a) 6   b) 5   c) 4   d) 3   e) 2   f) 1   g) $\frac{1}{2}$ h) $\frac{1}{4}$

$$u = \frac{x^2}{y} \quad v = \frac{y^2}{x}$$

$$\frac{u}{v} = \frac{x^3}{y^3}$$

$$y = x^2 \Rightarrow \frac{x^2}{y} = 1 \Rightarrow u = 1$$

$$y = \frac{1}{2}x^2 \Rightarrow \frac{x^2}{y} = 2 \Rightarrow u = 2$$

$$\frac{v}{u} = \frac{y^3}{x^3}$$

$$x = y^2 \Rightarrow \frac{y^2}{x} = 1 \Rightarrow v = 1$$

$$x = \frac{1}{2}y^2 \Rightarrow \frac{y^2}{x} = 2 \Rightarrow v = 2$$

$$uv = xy$$

$$J(x,y) = \begin{vmatrix} \frac{2x}{y} & \frac{-x^2}{y^2} \\ \frac{-y^2}{x^2} & \frac{2y}{x} \end{vmatrix} = 4 - 1 = 3$$

$$J(u,v) = \frac{1}{J(x,y)} = \frac{1}{3}$$

$$\int_1^2 \int_1^2 4v \cdot \frac{1}{3} \, dv \, du = \frac{4}{3} \int_1^2 \left(\frac{v^2}{2}\right)\Big|_1^2 \, du$$

$$= \frac{2}{3}(4-1) \int_1^2 du = 2(u)\Big|_1^2 = \boxed{2}$$

2.33) Evaluate the integral

$$\int_0^1 \int_0^{2-2x} \frac{(2x-y)^2}{2x+y}\,dy\,dx$$

a) 0

b) $\dfrac{1}{3}$

c) $\dfrac{4}{9}$

d) $\dfrac{2}{3}$

e) $\dfrac{3}{4}$

f) 1

2.33) Evaluate the integral

$$\int_0^1 \int_0^{2-2x} \frac{(2x-y)^2}{2x+y} \, dy \, dx$$

a) 0
b) $\frac{1}{3}$
c) $\frac{4}{9}$
d) $\frac{2}{3}$
e) $\frac{3}{4}$
f) 1

$0 \le y \le 2-2x$
$0 \le x \le 1$

$y = 2-2x$

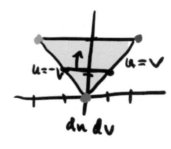

$J(x,y) = \begin{vmatrix} 2 & -1 \\ 2 & 1 \end{vmatrix} = 2--2 = 4$

$u = 2x - y$
$v = 2x + y$

$J(u,v) = \frac{1}{J(x,y)} = \frac{1}{4}$

$x\,y$	$\begin{array}{c}u=2x-y\\v=2x+y\end{array}$	$u\,v$
$(0,0)$	$\begin{array}{c}u=0\\v=0\end{array}$	$(0,0)$
$(1,0)$	$\begin{array}{c}u=2\\v=2\end{array}$	$(2,2)$
$(0,2)$	$\begin{array}{c}u=-2\\v=2\end{array}$	$(-2,2)$

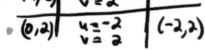

$u=-v$ $u=v$

$du\,dv$

$$\int_0^2 \int_{-v}^{v} \frac{u^2}{v} \cdot |J(u,v)| \, du \, dv = \frac{1}{4} \int_0^2 \frac{1}{v} \cdot \left(\frac{u^3}{3}\right)_{-v}^{v} dv$$

$$\frac{1}{v} \cdot \frac{1}{3}\underbrace{\left(v^3 - (-v)^3\right)}_{2v^3}$$

$$= \frac{1}{4} \cdot \frac{2}{3} \int_0^2 v^2 \, dv = \frac{1}{6}\left(\frac{v^3}{3}\right)_0^2 = \frac{1}{6}\left(\frac{8}{3} - 0\right) = \frac{8}{18} = \boxed{\frac{4}{9}}$$

2.34) Evaluate the integral

$$\iint\limits_R 4x^2\,dA$$

where R is the region in the first quadrant bounded by the graphs of

$$y = \frac{1}{x}, y = \frac{9}{x}, x = y, \text{ and } x = 16y$$

R has four corners $(1,1), (3,3), \left(12, \frac{3}{4}\right)$, and $\left(4, \frac{1}{4}\right)$.

Hint : $u = xy$ and $v = \dfrac{x}{y}$.

a) 60   e) 240
b) 120   f) 360
c) 180   g) 640
d) 225   h) 1200

2.34) Evaluate the integral

$$\iint\limits_R 4x^2 dA$$

where R is the region in the first quadrant bounded by the graphs of

$$y = \frac{1}{x}, y = \frac{9}{x}, x = y, \text{ and } x = 16y$$

R has four corners $(1,1), (3,3), \left(12, \frac{3}{4}\right)$, and $\left(4, \frac{1}{4}\right)$.

Hint : $u = xy$ and $v = \frac{x}{y}$.

a) 60   e) 240
b) 120   f) 360
c) 180   g) 640

d) 225   h) 1200

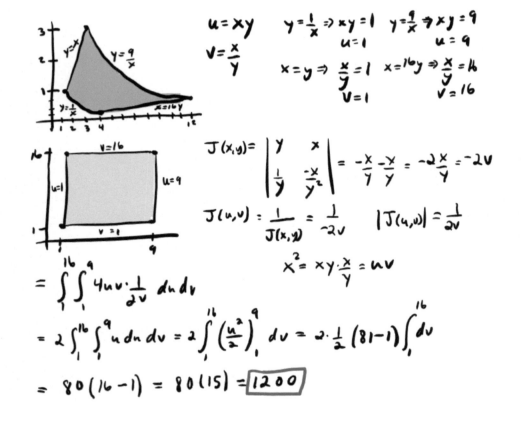

Section 3: Vector Calculus

3.1) Let

$$\mathbf{F} = \left\langle \frac{1}{\sqrt{x+y+z}}, \frac{1}{\sqrt{x+y+z}}, \frac{1}{\sqrt{x+y+z}} \right\rangle$$

Calculate

$$\int_C \mathbf{F} \cdot d\mathbf{r}$$

where C is the straight line path from $(1,-1,4)$ to $(5,-4,3)$

a) $\sqrt{5}$ b) $\sqrt{2}$ c) $\dfrac{1}{2}$ d) $\dfrac{1}{3}$ e) 2   f) 1   g) 0   h) None of these

3.1) Let

$$\mathbf{F} = \left\langle \frac{1}{\sqrt{x+y+z}}, \frac{1}{\sqrt{x+y+z}}, \frac{1}{\sqrt{x+y+z}} \right\rangle$$

Calculate

$$\int_C \mathbf{F} \cdot d\mathbf{r}$$

where C is the straight line path from $(1,-1,4)$ to $(5,-4,3)$

a) $\sqrt{5}$ b) $\sqrt{2}$ c) $\dfrac{1}{2}$ d) $\dfrac{1}{3}$ e) 2   f) 1   g) 0   h) None of these

$$(1,-1,4) \longrightarrow (5,-4,3)$$

$$\text{Pt} \qquad\qquad \text{Dir. vector}$$

$$X = 1 + (5-1)t \qquad x = 1+4t \qquad d_x = 4dt$$
$$y = -1 + (-4-1)t \qquad y = -1-3t \qquad dy = -3dt$$
$$z = 4 + (3-4)t \qquad z = 4-t \qquad dz = -dt$$

$$r = \langle 1+4t, -1-3t, 4-t \rangle \Rightarrow dr = \langle 4, -3, -1 \rangle dt$$

$$\text{rewrite } F_{onc} = \left\langle \tfrac{1}{2}, \tfrac{1}{2}, \tfrac{1}{2} \right\rangle = \tfrac{1}{2} \langle 1, 1, 1 \rangle$$

$$x+y+z = 4$$
$$\sqrt{x+y+z} = 2 \qquad\qquad F \cdot dr = \tfrac{1}{2}(4-3-1) = 0$$

$$\int F \cdot dr = \boxed{0}$$

3.2) Calculate $\int_C \mathbf{F} \cdot \mathbf{dr}$

where $\mathbf{F} = \langle z^2, x, y \rangle$ and C is the path

$\mathbf{r} = \langle 3 + 5t^2, 3 - t^2, t \rangle$, $0 \le t \le 2$

3.2) Calculate $\int_C \mathbf{F} \cdot \mathbf{dr}$

where $\mathbf{F} = \langle z^2, x, y \rangle$ and C is the path

$\mathbf{r} = \langle 3 + 5t^2, 3 - t^2, t \rangle$, $0 \le t \le 2$

Closed? No
Ind. of path? No

use given parametrization

$\rightarrow$ Close the path and use Green's Thm.

For C:
$\mathbf{F} = \langle t^2, 3 + 5t^2, 3 - t^2 \rangle$

$dr = \langle 10t, -2t, 1 \rangle dt$

$$\mathbf{F} \cdot dr = 10t^3 - 2t(3 + 5t^2) + 3 - t^2$$
$$= 10t^3 - 6t - 10t^3 + 3 - t^2$$

$\mathbf{F} \cdot dr = -6t + 3 - t^2$

$$\int_C \mathbf{F} \cdot dr = \int_0^2 (-t^2 - 6t + 3) dt$$

$$= \left[-\frac{t^3}{3} - 3t^2 + 3t \right]_0^2$$

$$= -\frac{8}{3} - 3(4) + 3(2) - 0$$

$$= -\frac{8}{3} - 12 + 6 = -\frac{8}{3} - 6 = \frac{-8 - 18}{3} = \boxed{\frac{-26}{3}}$$

3.3) Let

$$\mathbf{F} = \left\langle e^x \ln y, \ \frac{e^x}{y} + \sin z, \ y \cos z \right\rangle$$

Use the Fundamental Theorem of Line Integrals to evaluate

$\displaystyle\int_C \mathbf{F} \cdot d\mathbf{r}$ where C is any curve from $(7, 1, \pi)$ to $\left(1, e, \dfrac{\pi}{6}\right)$

a) $\dfrac{3e}{2}$ b) $\dfrac{e^2}{4}$ c) $\dfrac{e}{4}$ d) $\dfrac{1}{2}$ e) e f) $4 - \dfrac{e}{2}$ g) e^2 h) 1

3.3) Let

$$\mathbf{F} = \left\langle e^x \ln y, \ \frac{e^x}{y} + \sin z, \ y \cos z \right\rangle$$

Use the Fundamental Theorem of Line Integrals to evaluate

$$\int_C \mathbf{F} \cdot \mathbf{dr} \ \text{ where } C \text{ is any curve from } (7,1,\pi) \text{ to } \left(1, e, \frac{\pi}{6}\right)$$

a) $\dfrac{3e}{2}$ b) $\dfrac{e^2}{4}$ c) $\dfrac{e}{4}$ d) $\dfrac{1}{2}$ e) e f) $4 - \dfrac{e}{2}$ g) e^2 h) 1

$P = e^x \ln y \qquad P_y = e^x \cdot \frac{1}{y} \overset{?\checkmark}{=} Q_x = \frac{e^x}{y}$

$Q = \frac{e^x}{y} + \sin z \qquad P_z = 0 \overset{?\checkmark}{=} R_x = 0$

$R = y \cos z \qquad Q_z = \cos z \overset{?\checkmark}{=} R_y = \cos z$

$\int F \cdot dr$ is independent of path. Use FTLI $\Rightarrow$ find ϕ

① $\phi = \int P \, dx = \int (e^x \ln y) \, dx = e^x \ln y + G(y, z)$

② $\phi = \int Q \, dy = \int (\frac{e^x}{y} + \sin z) \, dy = e^x \cdot \ln y + y \cdot \sin z + H(x, z)$

③ $\phi = \int R \, dz = \int (y \cos z) \, dz = y \sin z + K(x, y)$

①② $\Rightarrow$ $e^x \ln y + G(y, z) = e^x \ln y + y \sin z + H(x, z)$

$\qquad\qquad G(y, z) = y \sin z + H(x, z) \leftarrow$ can't have x's

$\qquad\qquad G(y, z) = y \sin z + H(z)$

②③ $\Rightarrow$ $e^x \ln y + y \sin z + H(z) = y \sin z + K(x, y)$

$\qquad\qquad e^x \ln y + H(z) = K(x, y)$

$\qquad\qquad\qquad\qquad$ can't have z's so it is just a constant

$\phi(x,y,z) = e^x \ln y + y \sin z + C$

$\int F \cdot dr = \phi(1, e, \frac{\pi}{6}) - \phi(7, 1, \pi)$

$\qquad = (e \cdot \ln e + e \sin \frac{\pi}{6}) - (e^7 \ln 1 + 1 \cdot \sin \pi)$

$\qquad = e + e \cdot \frac{1}{2} - 0 - 0 = \boxed{\dfrac{3e}{2}}$

3.4) Calculate $\int_C \mathbf{F} \cdot \mathbf{dr}$

where $\mathbf{F} = \left\langle 2x + e^{-y}, 4y - xe^{-y} \right\rangle$

and C goes from $(0,0)$ to $(1,1)$ along $y = x^4$.

3.4) Calculate $\int_C \mathbf{F} \cdot \mathbf{dr}$

where $\mathbf{F} = \langle 2x + e^{-y}, 4y - xe^{-y} \rangle$

and C goes from $(0,0)$ to $(1,1)$ along $y = x^4$.

Closed? No
Ind. of path? Yes

$Q_x = -e^{-y}$

$P_y = -e^{-y}$ ✓

Pick a convenient path

Find the potential function ϕ with $\mathbf{F} = \nabla \phi$

$\phi = \int (2x + e^{-y}) dx = x^2 + xe^{-y} + G(y)$

$\phi = \int (4y - xe^{-y}) dy = 2y^2 + xe^{-y} + H(x)$

$x^2 + \cancel{xe^{-y}} + G(y) = 2y^2 + \cancel{xe^{-y}} + H(x)$

$\Rightarrow \quad H(x) = x^2 + c \quad$ and $\quad G(y) = 2y^2 + c$

$\phi = x^2 + 2y^2 + xe^{-y} + c$

FTLI $\Rightarrow$ Answer $= \phi(1,1) - \phi(0,0)$

$\phi(1,1) = 1 + 2 + e^{-1}$

$- \phi(0,0) = 0 + 0 + 0$

$= \boxed{3 + \dfrac{1}{e}}$

3.5) Calculate $\int\limits_{C} \mathbf{F} \cdot \mathbf{dr}$

where $\mathbf{F} = \left\langle y + yz, x + 3z^3 + xz, 9yz^2 + xy - 1 \right\rangle$

and C goes from $(1,1,1,)$ to $(2,1,4)$ along any path.

3.5) Calculate $\int_C \mathbf{F} \cdot \mathbf{dr}$

where $\mathbf{F} = \left\langle y + yz, x + 3z^3 + xz, 9yz^2 + xy - 1 \right\rangle$

and C goes from $(1,1,1,)$ to $(2,1,4)$ along any path.

$$P = y + yz \qquad P_y = 1+z \overset{?\checkmark}{=} Q_x = 1+z$$

$$Q = x + 3z^3 + xz \qquad P_z = y \overset{?\checkmark}{=} R_x = y$$

$$R = 9yz^2 + xy - 1 \qquad Q_z = 9z^2 + x \overset{?\checkmark}{=} R_y = 9z^2 + x$$

$\int \mathbf{F} \cdot \mathbf{dr}$ is independent of path. $\rightarrow$ use FTLI, Find ϕ

$\rightarrow$ use straight line path

① $\phi = \int P\,dx = \int (y + yz)\,dx = xy + xyz + G_1(y,z)$

② $\phi = \int Q\,dy = \int (x + 3z^3 + xz)\,dy = xy + 3z^3 y + xyz + H(x,z)$

③ $\phi = \int R\,dz = \int (9yz^2 + xy - 1)\,dz = 3z^3 y + xyz - z + K(x,y)$

①② $\Rightarrow$ $\cancel{xy} + \cancel{xyz} + G(y,z) = \cancel{xy} + 3z^3 y + \cancel{xyz} + H(x,z)$

$\qquad\qquad G(y,z) = 3z^3 y + \underbrace{H(x,z)}_{H(z)}$ can't have x's so $H(z)$

②③ $\qquad \cancel{xy} + \cancel{xyz} + 3z^3 y + H(z) = 3z^3 y + \cancel{xyz} - z + K(x,y)$

$\qquad\qquad xy + 1 + (z) = -z + K(x,y)$

$\qquad\qquad \Rightarrow xy = K(x,y) \,\&\, H(z) = -z$ (they each can have $+C$)

$\phi(x,y,z) = xy + xyz + 3z^3 y - z + C$

$\int \mathbf{F} \cdot \mathbf{dr} = \phi(2,1,4) - \phi(1,1,1) = (2 + 8 + 64 \cdot 3 - 4) - (1 + 1 + 3 - 1)$

$\qquad = 10 + 192 - 4 - 4 = \boxed{194}$

3.6) Find the work done by the force field

$$\mathbf{F} = e^{y}\,\mathbf{i} + \left(xe^{y} + e^{z}\right)\mathbf{j} + ye^{z}\mathbf{k}$$

on a particle as it moves along the helix given by

$$\mathbf{r}(t) = \cos t\,\mathbf{i} + \sin t\,\mathbf{j} + 2t\,\mathbf{k}$$

from the point $(1,0,0)$ to the point $(0,1,\pi)$.

a) e^{π} b) $e^{\pi} - 2$ c) $e^{\pi} + 1$ d) $e^{\pi} - 1$ e) $2e^{\pi} - 1$ f) $2e^{\pi} - 3$

3.6) Find the work done by the force field

$$\mathbf{F} = e^y \, \mathbf{i} + \left(xe^y + e^z\right) \mathbf{j} + ye^z \mathbf{k}$$

on a particle as it moves along the helix given by

$$\mathbf{r}(t) = \cos t \, \mathbf{i} + \sin t \, \mathbf{j} + 2t \, \mathbf{k}$$

from the point $(1,0,0)$ to the point $(0,1,\pi)$.

a) e^π b) $e^\pi - 2$ c) $e^\pi + 1$ d) $e^\pi - 1$ e) $2e^\pi - 1$ f) $2e^\pi - 3$

$\int \mathbf{F} \cdot d\mathbf{r}$

Closed? No

Ind. of path? 3D

(any path, straight line)

FTLT find ϕ

$P_y = e^y \overset{?}{=} Q_x = e^y$

$P_z = 0 \overset{?\sqrt{}}{=} R_x = 0$

$Q_z = e^z \overset{?\sqrt{}}{=} R_y = e^z$

$\phi = \int P \, dx = \int e^y \, dx = xe^y + G(y,z)$

$\phi = \int Q \, dy = \int (xe^y + e^z) \, dy = xe^y + ye^z + H(x,z)$

$\phi = \int R \, dz = \int ye^z \, dz = ye^z + K(x,y)$

$\phi = xe^y + ye^z + C$

$\int \mathbf{F} \cdot d\mathbf{r} = \phi(0,1,\pi) - \phi(1,0,0)$

$= (0 + e^\pi) - (1 + 0) = \boxed{e^\pi - 1}$

3.7) Calculate

$$\int_C (8x \sin y)\,dx + (-8y \cos x)\,dy$$

where C is the rectangle cut from the first quadrant

by the lines $y = \dfrac{\pi}{6}$ and $x = \dfrac{\pi}{3}$ traversed clockwise.

a) $\dfrac{\pi^2}{6}$ b) $\dfrac{\pi^2}{9}$ c) $\dfrac{\pi^2}{3}$ d) $\dfrac{\pi^2}{18}$ e) $\dfrac{5\pi^2}{18}$ f) $\dfrac{\pi^2}{4}$ g) $\dfrac{\pi^2}{16}$ h) 0

3.7) Calculate

$$\int_C (8x \sin y)\,dx + (-8y\cos x)\,dy$$

where C is the rectangle cut from the first quadrant

by the lines $y = \dfrac{\pi}{6}$ and $x = \dfrac{\pi}{3}$ traversed clockwise.

a) $\dfrac{\pi^2}{6}$ b) $\dfrac{\pi^2}{9}$ c) $\dfrac{\pi^2}{3}$ d) $\dfrac{\pi^2}{18}$ e) $\dfrac{5\pi^2}{18}$ f) $\dfrac{\pi^2}{4}$ g) $\dfrac{\pi^2}{16}$ h) 0

$x = \frac{\pi}{3}$

$y = \frac{\pi}{6}$

C

R

Clockwise

closed? Yes

$\oint F \cdot dr$
Indep.
of path? No

$P_y = 8x\cos y$
$Q_x = 8 y \sin x$ $P_y \neq Q_x$

Parametrize C

⟶ use Green's
Theorem ✓

$$\oint_C F \cdot dr = \iint_R (Q_x - P_y)\,dA$$

counter-clockwise

$$-\oint F \cdot dr = -\iint_R (Q_x - P_y)\,dA = -\int_0^{\pi/3}\int_0^{\pi/6} 8(y\sin x - x\cos y)\,dy\,dx$$

$$= -8\int_0^{\pi/3}\left[\frac{y^2}{2}\sin x - x\sin y\right]_0^{\pi/6}\,dx = -8\int_0^{\frac{\pi}{3}}\left[\left(\frac{\pi^2}{72}\sin x - x\cdot\frac{1}{2}\right) - 0\right]dx$$

$$= -8\cdot\left[-\frac{\pi^2}{72}\cos x - \frac{x^2}{4}\right]_0^{\pi/3} = -8\left[\left(-\frac{\pi^2}{72}\left(\frac{1}{2}\right) - \frac{\pi^2}{36}\right) - 0\right]$$

$$= \frac{\pi^2}{18} + \frac{2\pi^2}{18} = \frac{3\pi^2}{18} = \boxed{\dfrac{\pi^2}{6}}$$

3.8) Calculate

$$\oint_C xy\,dx + \left(x^2 + x\right)dy$$

where C is the path given.

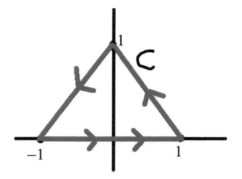

3.8) Calculate

$$\oint_C xy\,dx + \left(x^2 + x\right)dy$$

where C is the path given.

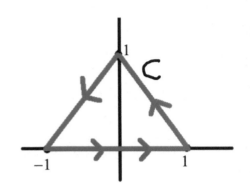

$x = y - 1$ 　　 $x = 1 - y$
$y = x + 1$ 　　 $y = 1 - x$

$\iint dy\,dx$ need 2 int

$\iint dx\,dy$ ✓

Closed? Yes
Indep. of path? No $\Big\}$ use Green's Thm.

$Q_x = 2x + 1$
$P_y = x$ 　　 $Q_x - P_y = 2x + 1 - x = x + 1$

$$\oint_C F\,dr = \iint_R (Q_x - P_y)\,dA = \int_0^1 \int_{y-1}^{1-y} (x+1)\,dx\,dy$$

$$= \int_0^1 \left(\frac{x^2}{2} + x\right)\Big|_{y-1}^{1-y} dy = \int_0^1 \left[\frac{(1-y)^2}{2} + 1 - y\right] - \left[\frac{(y-1)^2}{2} + y - 1\right] dy$$

$$\frac{1 - 2y + y^2}{2} + 1 - y - \left[\frac{y^2 - 2y + 1}{2} + y - 1\right]$$

$$= \int_0^1 \frac{1}{2} - y + \frac{y^2}{2} + 1 - y - \frac{y^2}{2} + \cancel{y} - \frac{1}{2} - \cancel{y} + 1$$

$$= \int_0^1 (2 - 2y)\,dy = (2y - y^2)\Big|_0^1 = (2 - 1) - 0 = \boxed{1}$$

3.9) Calculate

$$\int_C \left(-16y + \sin\left(x^2\right)\right)dx + \left(4e^y + 3x^2\right)dy$$

where C is the path pictured below.

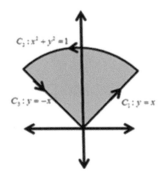

a) $\dfrac{7\pi}{2}$ b) 3π c) $\dfrac{15\pi}{4}$ d) $\dfrac{12\pi}{5}$

e) $\dfrac{13\pi}{3}$ f) $\dfrac{14\pi}{5}$ g) 4π h) $\dfrac{16\pi}{5}$

3.9) Calculate

$$\int_C \left(-16y + \sin\left(x^2\right)\right)dx + \left(4e^y + 3x^2\right)dy$$

where C is the path pictured below.

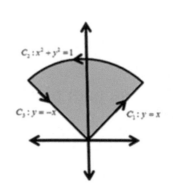

a) $\dfrac{7\pi}{2}$ b) 3π c) $\dfrac{15\pi}{4}$ d) $\dfrac{12\pi}{5}$

e) $\dfrac{13\pi}{3}$ f) $\dfrac{14\pi}{5}$ g) 4π h) $\dfrac{16\pi}{5}$

Closed? Yes $\left\{\begin{array}{l} \text{Parametrize the given path} \\ \\ \text{use Green's Thm} \end{array}\right.$

$\left.\begin{array}{l}\text{Indep. of Path?} \\ P_y = -16 \\ Q_x = 6x \end{array}\right\}$ No

$$\oint_C F \cdot dr = \iint_R (Q_x - P_y)\,dA = \iint_R (6x - {}^-16)\,dA = \iint_R (6x + 16)\,dA$$

R is "circular" in nature, we Polar

$0 \le r \le 1$
$\frac{\pi}{4} \le \theta \le \frac{3\pi}{4}$

since $y=x$ since $y=-x$

$$= \int_{\pi/4}^{3\pi/4} \int_0^1 (6r\cos\theta + 16)\,r\,dr\,d\theta$$

$$= \int_{\pi/4}^{3\pi/4} \int_0^1 (6r^2\cos\theta + 16r)\,dr\,d\theta = \int_{\frac{\pi}{4}}^{3\pi/4} \left[2r^3\cos\theta + 8r^2\right]_0^1 d\theta$$

$$= \int_{\pi/4}^{3\pi/4} (2\cos\theta + 8)\,d\theta = \left[2\sin\theta + 8\theta\right]_{\pi/4}^{3\pi/4}$$

$$= \left(2\left(\tfrac{\sqrt{2}}{2}\right) + 8\cdot\tfrac{3\pi}{4}\right) - \left(2\left(\tfrac{\sqrt{2}}{2}\right) + 8\cdot\tfrac{\pi}{4}\right) = \sqrt{2} + 6\pi - \sqrt{2} - 2\pi$$

$$= \boxed{4\pi}$$

3.10) Calculate $\displaystyle\int_C \mathbf{F} \cdot \mathbf{dr}$

where $\mathbf{F} = \left\langle 6x^2 e^{2x^3 - 2y^3}, -6y^2 e^{2x^3 - 2y^3} \right\rangle$

and C is the semicircle $x^2 + y^2 = 1$, $y \geq 0$,

traversed from $(-1, 0)$ to $(1, 0)$.

3.10) Calculate $\int_C \mathbf{F} \cdot \mathbf{dr}$

where $\mathbf{F} = \left\langle 6x^2 e^{2x^3 - 2y^3}, -6y^2 e^{2x^3 - 2y^3} \right\rangle$

and C is the semicircle $x^2 + y^2 = 1$, $y \geq 0$,

traversed from $(-1, 0)$ to $(1, 0)$.

$Q_x = -6y^2 e^{2x^3 - 2y^3} \cdot (6x^2)$

$P_y = 6x^2 e^{2x^3 - 2y^3} \cdot (-6y^2)$

$\left. \begin{array}{l} \end{array} \right\} Q_x = P_y \Rightarrow \text{Ind. of Path}$

Not closed

$\left. \begin{array}{l} \end{array} \right\}$ ⟶ Pick a conv. path ✓

⟶ Find ϕ and use FTLI

$C_1: \begin{array}{l} x = t \\ y = 0 \end{array} \quad -1 \leq t \leq 1$

$\vec{r} = \langle t, 0 \rangle$

$\mathbf{dr} = \langle 1, 0 \rangle \, dt$

$F_{on\, C_1} = \langle 6t^2 e^{2t^3}, 0 \rangle$

$\mathbf{F} \cdot \mathbf{dr} = 6t^2 e^{2t^3} \, dt$

$u = 2t^3$

$du = 6t^2 \, dt \quad \int e^u \, du = e^u$

$\int_C \mathbf{F} \cdot \mathbf{dr} = \int_{C_1} \mathbf{F} \cdot \mathbf{dr}$

$= \int_{-1}^{1} 6t^2 e^{2t^3} \, dt = \left[e^{2t^3} \right]_{-1}^{1} = \boxed{e^2 - e^{-2}}$

3.11) Calculate $\int_C \mathbf{F}\cdot\mathbf{dr}$

where $\mathbf{F} = \langle y\cos x - xy\sin x,\ xy + x\cos x\rangle$

and C is the triangle from $(0,0)$ to $(2,0)$ to $(0,4)$ to $(0,0)$.

a) $\dfrac{17}{6}$ b) $\dfrac{16}{3}$ c) $\dfrac{19}{3}$ d) $\dfrac{15}{4}$

e) $\dfrac{13}{4}$ f) $\dfrac{22}{3}$ g) $\dfrac{19}{4}$ h) None of these

3.11) Calculate $\int_C \mathbf{F} \cdot \mathbf{dr}$

where $\mathbf{F} = \langle y\cos x - xy\sin x, \, xy + x\cos x \rangle$

and C is the triangle from $(0,0)$ to $(2,0)$ to$(0,4)$ to$(0,0)$.

a) $\dfrac{17}{6}$ b) $\dfrac{16}{3}$ c) $\dfrac{19}{3}$ d) $\dfrac{15}{4}$

e) $\dfrac{13}{4}$ f) $\dfrac{22}{3}$ g) $\dfrac{19}{4}$ h) None of these

Closed? yes
Ind. of Path? No $\Big\}$ ← Parametrize given path
→ use Green's Thm

$P = y\cos x - xy\sin x \qquad P_y = \cos x - x\sin x$

$Q = xy + x\cos x \qquad Q_x = y + 1\cdot\cos x + x\cdot -\sin x$

$Q_x - P_y = (y + \cos x - x\sin x) - (\cos x - x\sin x)$

$Q_x - P_y = y$

Closed ✓
Counter
-clockwve ✓ $\Big\}$ use Green's Thm

$\displaystyle\oint_C \mathbf{F} \cdot d\mathbf{r} = \iint_R (Q_x - P_y)\, dA$

$n = \dfrac{-4}{2} = -2 \Big\} \; y = -2x + 4$
$b = 4$
$y = 4 - 2x$

$(4-2x)(4-2x)$
$2(2-x)\cdot 2(2-x)$
$4(2-x)(2-x)$
$4(4 - 4x + x^2)$

$\displaystyle = \int_0^2 \int_0^{4-2x} y \, dy\, dx$

$\displaystyle = \int_0^2 \left[\frac{y^2}{2}\right]_0^{4-2x} dx$

$\displaystyle = \int_0^2 \frac{4(4 - 4x + x^2)}{2}\, dx$

$\displaystyle = 2\int_0^2 (4 - 4x + x^2)\, dx = 2\left[4x - 2x^2 + \frac{x^3}{3}\right]_0^2$

$\displaystyle = 2\left[8 - 8 + \frac{8}{3}\right] = 2\cdot\frac{8}{3} = \boxed{\dfrac{16}{3}}$

3.12) Calculate $\int_C \mathbf{F} \cdot \mathbf{dr}$

where $\mathbf{F} = \left\langle 2xz + y^2, 2xy, x^2 + 6z^2 \right\rangle$

and C is the path $\mathbf{r} = \left\langle t^2, t+3, 4t-1 \right\rangle$, $0 \le t \le 1$

a) 45   b) 48   c) 50   d) 52
e) 56   f) 64   g) 75   h) None of these

3.12) Calculate $\int_C \mathbf{F} \cdot d\mathbf{r}$

where $\mathbf{F} = \langle 2xz + y^2, 2xy, x^2 + 6z^2 \rangle$

and C is the path $\mathbf{r} = \langle t^2, t+3, 4t-1 \rangle$, $0 \le t \le 1$

a) 45   b) 48   c) 50   d) 52
e) 56   f) 64   g) 75   h) None of these

Closed? (No)

$r(0) = \langle 0, 3, -1 \rangle$ st @ $(0, 3, -1)$
$r(1) = \langle 1, 4, 3 \rangle$ end @ $(1, 4, 3)$

Indep. of Path? (yes)

$P_y = 2y \checkmark Q_x = 2y$
$P_z = 2x \checkmark R_x = 2x$
$Q_z = 0 \checkmark R_y = 0$

Pick a conv.
(straight line)
path

use FTLI
Find ϕ

___

① $\phi = \int P\, dx = \int (2xz + y^2)\, dx = x^2 z + xy^2 + G(y, z)$

② $\phi = \int Q\, dy = \int (2xy)\, dy = xy^2 + H(x, z)$

③ $\phi = \int R\, dz = \int (x^2 + 6z^2)\, dz = x^2 z + 2z^3 + K(x, y)$

① & ② $\Rightarrow$ $x^2 z + \cancel{xy^2} + G(y, z) = \cancel{xy^2} + H(x, z)$

$x^2 z + G(y, z) = H(x, z)$

K can't have
y's

$\Rightarrow \phi = x^2 z + xy^2 + G(z)$

① & ③ $\Rightarrow$ $\cancel{x^2 z} + xy^2 + G(z) = \cancel{x^2 z} + 2z^3 + K(x, y)$

$G(z) = 2z^3 + K(x, y)$

must be a constant
since LHS only has z's

$\Rightarrow G(z) = 2z^3 + C$

$\phi = x^2 z + xy^2 + 2z^3 + C$

$\int_C F \cdot dr = \phi(1, 4, 3) - \phi(0, 3, -1) = (3 + 16 + 54) - (-2)$

$= 73 + 2 = \boxed{75}$

3.13) Evaluate $\int_C xy\,dx + (x-y)\,dy$

C consists of line segments from

$(0,0)$ to $(2,0)$ and from $(2,0)$ to $(3,2)$.

a) $\dfrac{8}{3}$ b) $\dfrac{17}{3}$ c) $\dfrac{22}{3}$ d) $\dfrac{4}{3}$

e) $\dfrac{2}{3}$ f) $\dfrac{26}{3}$ g) π h) None of these

3.13) Evaluate $\int_C xy\,dx + (x-y)\,dy$

C consists of line segments from $(0,0)$ to $(2,0)$ and from $(2,0)$ to $(3,2)$.

a) $\dfrac{8}{3}$ b) $\dfrac{17}{3}$ c) $\dfrac{22}{3}$ d) $\dfrac{4}{3}$

e) $\dfrac{2}{3}$ f) $\dfrac{26}{3}$ g) π h) None of these

Closed? No
Ind. of Path? No
$P_y = x$ $Q_x = 1$

$\left.\begin{array}{l}\end{array}\right\}$ → Parametrize the given path ✓
→ close the path and use Green's Theorem

$\int_C F \cdot dr = \int_{C_1} F \cdot dr + \int_{C_2} F \cdot dr$

C_1: $x = t$ $y = 0$ $0 \le t \le 2$ $r = \langle t, 0 \rangle$ $dr = \langle 1, 0 \rangle dt$ rewrite F on C_1: $F = \langle 0, t \rangle$

$F \cdot dr = 0 + 0$ so $\int_{C_1} F \cdot dr = 0$

C_2: $m = \dfrac{+2}{1} = 2$ $y = mx + b$ $0 = 2(2) + b$ $b = -4$ ⟹ $y = 2x - 4$

$x = t$
$y = 2t - 4$ $2 \le t \le 3$ $r = \langle t, 2t-4 \rangle$ $dr = \langle 1, 2 \rangle dt$ rewrite F on C_2: $F = \langle 2t^2 - 4t, 4-t \rangle$

$xy = t(2t-4) = 2t^2 - 4t$
$x - y = t - (2t-4) = 4 - t$

$F \cdot dr = (2t^2 - 4t)(1) + (4-t)(2) = 2t^2 - 4t + 8 - 2t = 2t^2 - 6t + 8$

$\int_2^3 (2t^2 - 6t + 8)\,dt = \left(2\dfrac{t^3}{3} - 3t^2 + 8t \right)\Big|_2^3 = (2 \cdot 9 - 27 + 24) - \left(\dfrac{16}{3} - 12 + 16\right)$

$= \underbrace{18 - 27 + 24}_{-9} - \dfrac{16}{3} + \underbrace{12 - 16}_{-4} = 11 - \dfrac{16}{3} = \dfrac{33-16}{3} = \boxed{\dfrac{17}{3}}$

3.14) Calculate $\displaystyle\int_C \mathbf{F}\cdot d\mathbf{r}$

where $\mathbf{F} = \left\langle \dfrac{z}{x}, \dfrac{z}{y}, \ln(xy) \right\rangle$

and C is the path $\mathbf{r} = \left\langle e^t, e^{2t}, t^2 \right\rangle$, $1 \le t \le 3$

A) 60   B) 64   C) 72   D) 75

E) 78   F) 81   G) 84   H) None of these

3.14) Calculate $\int_C \mathbf{F} \cdot \mathbf{dr}$

where $\mathbf{F} = \left\langle \dfrac{z}{x}, \dfrac{z}{y}, \ln(xy) \right\rangle$

and C is the path $\mathbf{r} = \left\langle e^t, e^{2t}, t^2 \right\rangle$, $1 \le t \le 3$

A) 60   B) 64   C) 72   D) 75
E) 78   F) 81   G) 84   H) None of these

Closed? **NO**

$r(1) = \langle e, e^2, 1 \rangle$ st @ $(e, e^2, 1)$
$r(3) = \langle e^3, e^6, 9 \rangle$ end @ $(e^3, e^6, 9)$

Ind. of Path? **Yes**

$P_y = 0 \checkmark Q_x = 0$
$P_z = \frac{1}{x} \checkmark R_x = \frac{1 \cdot y}{xy} = \frac{1}{x}$
$Q_z = \frac{1}{y} \checkmark R_y = \frac{1}{xy} x = \frac{1}{y}$

Pick a conv. (straight line) path

use FTLI
Find Φ ✓

① $\Phi = \int P\,dx = \int \frac{z}{x}\,dx = z \cdot \ln x + G(y,z)$

② $\Phi = \int Q\,dy = \int \frac{z}{y}\,dy = z \cdot \ln y + H(x,z)$

③ $\Phi = \int R\,dz = \int \ln(xy)\,dz = z \cdot \ln(xy) + K(x,y)$
$\qquad = z \cdot \left[\ln x + \ln y\right] + K(x,y)$
$\qquad = z \ln x + z \ln y + F(x,y)$

①&② ⟹ $z \ln x + G(y,z) = z \ln y + H(x,z)$ No info at this moment

①&③ ⟹ $z \ln x + G(y,z) = z \ln x + z \ln y + K(x,y)$
$\qquad\qquad G(y,z) = z \ln y + K(x,y)$ can't have x's
$\qquad\qquad G(y,z) = z \ln y + K(y)$
$\qquad\quad \Phi = z \ln x + z \ln y + K(y)$

②&③ $z \ln y + H(x,z) = z \ln x + z \ln y + K(y)$
$\qquad\qquad H(x,z) = z \ln x + K(y)$
$\qquad\qquad$ must be a constant since LHS H(x,z)

○

$\Phi = z \ln x + z \ln y + C$

$\int_C F \cdot dr = \Phi(e^3, e^6, 9) - \Phi(e, e^2, 1) = (9 \cdot \ln e^3 + 9 \ln e^6) - (\ln e + \ln e^2)$

$\qquad = 9 \cdot 3 + 9 \cdot 6 - 1 - 2 = 81 - 3 = \boxed{78}$

3.15) Green's theorem for area states that the region whose boundary is a simple closed curve satisfies

$$\text{Area of } R = \frac{1}{2} \oint_C x\,dy - y\,dx.$$

Find the area of the region enclosed by one arc of the cycloid $x = a(t - \sin t)$, $y = a(1 - \cos t)$ $0 \le t \le 2\pi$ and the x − axis.

A) $\dfrac{7\pi}{6}$ B) $\dfrac{11\pi}{12}$ C) $\dfrac{5\pi}{4}$ D) $\dfrac{5\pi}{12}$

E) $6\pi a^2$ F) $\dfrac{5\pi}{2}$ G) $\sqrt{2\pi}$ H) None of these

3.15) Green's theorem for area states that the region whose boundary is a simple closed curve satisfies

$$\text{Area of } R = \frac{1}{2}\oint_C x\,dy - y\,dx.$$

Find the area of the region enclosed by one arc of the cycloid $x = a(t - \sin t)$, $y = a(1 - \cos t)$ $0 \le t \le 2\pi$ and the x-axis.

A) $\dfrac{7\pi}{6}$ B) $\dfrac{11\pi}{12}$ C) $\dfrac{5\pi}{4}$ D) $\dfrac{5\pi}{12}$

E) $6\pi a^2$ F) $\dfrac{5\pi}{2}$ G) $\sqrt{2}\pi$ H) None of these

$$\frac{1}{2}\oint_C x\,dy - y\,dx = \frac{1}{2}\left[\int_{C_1} x\,dy - y\,dx + \int_{C_2} x\,dy - y\,dx \right]$$

C_2: $x = a(t - \sin t)$ $y = a(1 - \cos t)$
        $dx = a(1 - \cos t)$ $dy = a(\sin t)$

$x\,dy = a^2 (t\sin t - \sin^2 t)$
$-y\,dx = a^2(1 - 2\cos t + \cos^2 t)$
_____
$= a^2\left(t\sin t \underbrace{-\sin^2 t - \cos^2 t}_{-1} - 1 + 2\cos t \right)$

$x\,dy - y\,dx = a^2\left(t\sin t - 2 + 2\cos t \right)$

$\int_{C_1} x\,dy - y\,dx = \int_0^{2\pi} (a^2 t\sin t - 2 + 2\cos t)\,dt$

$= -a^2\left[-t\cos t + \sin t - 2t + 2\sin t \right]\Big|_0^{2\pi}$

$= -a^2\left[(t \cdot 2\pi + 0 - 4\pi + 0) - (0 + 0 + 0) \right]$

$= -a^2\left[-6\pi \right] = \boxed{6\pi a^2}$

$\frac{1}{2}\oint_C x\,dy - y\,dx = \frac{1}{2}\left[0 + 6\pi a^2 \right] = \boxed{3\pi a^2}$

Must be counter-clockwise

C_1: $x = t$ $dx = 1$
        $y = 0$ $dy = 0$
$x\,dy - y\,dx = 0 - 0 = 0$

$\int_{C_1} x\,dy - y\,dx = \boxed{0}$

D   I
$\sin t$
$-\cos t$
$-\sin t$

3.16) Find the curl of the vector field

$$\mathbf{F} = z^2 y \cos(xy) \, \mathbf{i} + z^2 x (1 + \cos(xy)) \, \mathbf{j} + 2z \sin(xy) \mathbf{k}$$

3.16) Find the curl of the vector field

$$\mathbf{F} = z^2 y \cos(xy)\ \mathbf{i} + z^2 x\left(1 + \cos(xy)\right)\ \mathbf{j} + 2z\sin(xy)\mathbf{k}$$

$$\text{Curl } \vec{F} = \nabla \times F = \begin{vmatrix} \mathbf{i} & \mathbf{j} & \mathbf{K} \\ \dfrac{\partial}{\partial x} & \dfrac{\partial}{\partial y} & \dfrac{\partial}{\partial z} \\ z^2 y\cos(xy) & z^2 x(1+\cos(xy)) & 2z\sin(xy) \end{vmatrix}$$

$$= \left\langle \dfrac{\partial}{\partial y}(2z\sin(xy)) - \dfrac{\partial}{\partial z}\left(z^2 x(1+\cos(xy))\right), -\left[\dfrac{\partial}{\partial x}(2z\sin(xy)) - \dfrac{\partial}{\partial z}(z^2 y\cos(xy))\right], \dfrac{\partial}{\partial x}\left(z^2 x(1+\cos(xy))\right) - \dfrac{\partial}{\partial y}(z^2 y\cos(xy)) \right\rangle$$

$$= \left\langle 2zx\cos(xy) - 2z\,x\,(1+\cos(xy)), -\left[2zy\cos(xy) - 2zy\cos(xy)\right], \right.$$

$$\underbrace{}_{0}$$

$$\left. z^2(1+\cos(xy)) + z^2 x(-y\sin(xy)) - \left[z^2\cos(xy) - z^2 y \times \sin(xy)\right] \right\rangle$$

$$= \boxed{\left\langle -2zx\ ,\ \ 0\ \ ,\ z^2 \right\rangle}$$

3.17) Let

$$\mathbf{F} = e^{-x}\cos(y)\ \mathbf{i} + e^{-x}\sin y\ \mathbf{j} + \ln z\ \mathbf{k}$$

Find the div $\mathbf{F}$ and the curl $\mathbf{F}$.

3.17) Let

$$\mathbf{F} = e^{-x}\cos(y)\ \mathbf{i} + e^{-x}\sin y\ \mathbf{j} + \ln z\ \mathbf{k}$$

Find the div $\mathbf{F}$ and the curl $\mathbf{F}$.

$$\text{div}\,\mathbf{F} = \nabla \cdot \mathbf{F} = \left\langle \tfrac{\partial}{\partial x}, \tfrac{\partial}{\partial y}, \tfrac{\partial}{\partial z} \right\rangle \cdot \left\langle e^{-x}\cos y, e^{-x}\sin y, \ln z \right\rangle$$

$$\text{div}\,\mathbf{F} = \tfrac{\partial}{\partial x}(e^{-x}\cos y) + \tfrac{\partial}{\partial y}(e^{-x}\sin y) + \tfrac{\partial}{\partial z}(\ln z)$$

$$\text{div}\,\mathbf{F} = -e^{-x}\cos y + e^{-x}\cos y + \tfrac{1}{z} = \boxed{\tfrac{1}{z}}$$

$$\text{curl}\,\mathbf{F} = \nabla \times \mathbf{F} = \begin{vmatrix} \mathbf{i} & \mathbf{j} & \mathbf{k} \\ \tfrac{\partial}{\partial x} & \tfrac{\partial}{\partial y} & \tfrac{\partial}{\partial z} \\ e^{-x}\cos y & e^{-x}\sin y & \ln z \end{vmatrix}$$

$$= \left\langle \tfrac{\partial}{\partial y}(\ln z) - \tfrac{\partial}{\partial z}(e^{-x}\sin y),\ -\left[\tfrac{\partial}{\partial x}(\ln z) - \tfrac{\partial}{\partial z}(e^{-x}\cos y)\right]\tfrac{\partial}{\partial z}(e^{-x}\sin y) - \tfrac{\partial}{\partial y}(e^{-x}\cos y) \right\rangle$$

$$= \left\langle 0-0,\ -[0-0],\ e^{-x}\sin y - -1 e^{-x}\sin y \right\rangle = \boxed{\langle 0, 0, 0 \rangle}$$

3.18) The surface of the dome on a building is given by

$$\mathbf{r}(u,v) = 12\sin u \cos(v)\ \mathbf{i} + 12\sin u \sin v\ \mathbf{j} + 12\cos u\ \mathbf{k}$$

where $0 \le u \le \dfrac{\pi}{3}, 0 \le v \le 2\pi,$ and $\mathbf{r}$ is in meters.

Find the surface area of the dome.

3.18) The surface of the dome on a building is given by

$$\mathbf{r}(u,v) = 12\sin u \cos(v)\ \mathbf{i} + 12\sin u \sin v\ \mathbf{j} + 12\cos u\ \mathbf{k}$$

where $0 \le u \le \dfrac{\pi}{3}, 0 \le v \le 2\pi$, and $\mathbf{r}$ is in meters.

Find the surface area of the dome.

Surface Area $= \iint_S 1\, dS$ with S defined parametrically $\mathbf{r}(u,v)$

$$dS = \| r_u \times r_v \|\, du\, dv$$

$\mathbf{r}(u,v) = \langle 12\sin u \cos v,\ 12\sin u \sin v,\ 12\cos u \rangle$

$r_u = \langle 12\cos u \cos v,\ 12\cos u \sin v,\ -12\sin u \rangle$

$r_v = \langle -12\sin u \sin v,\ 12\sin u \cos v,\ 0 \rangle$

$$r_u \times r_v = \begin{vmatrix} i & j & k \\ 12\cos u \cos v & 12\cos u \sin v & -12\sin u \\ -12\sin u \sin v & 12\sin u \cos v & 0 \end{vmatrix}$$

$= \langle 144\sin^2 u \cos v,\ -144\sin^2 u \sin v,\ 144[\sin u \cos u \cos^2 v + \sin u \cos u \sin^2 v] \rangle$

$r_u \times r_v = 144 \langle \sin^2 u \cos v,\ -\sin^2 u \sin v,\ \sin u \cos u (\cos^2 v + \sin^2 v) \rangle$ (1)

$\| r_u \times r_v \| = 144 \sqrt{\sin^4 u \cos^2 v + \sin^4 u \sin^2 v + \sin^2 u \cos^2 u}$

$\| r_u \times r_v \| = 144 \sqrt{\sin^4 u (\cos^2 v + \sin^2 v) + \sin^2 u \cos^2 u}$ (1)

$\| r_u \times r_v \| = 144 \sqrt{\sin^2 u [\sin^2 u + \cos^2 u]} = 144\sin u$ (1)

Surface Area $= \iint_R 1 \cdot \| r_u \times r_v \|\, du\, dv = \int_0^{2\pi}\int_0^{\pi/3} 144\sin u\, du\, dv$

$= 144 \int_0^{2\pi} \underbrace{-[\cos u]_0^{\pi/3}}_{-(\frac{1}{2}-1)}\, dv = 144 \cdot \frac{1}{2} \underbrace{\int_0^{2\pi} dv}_{2\pi} = \boxed{144\pi\ \text{m}^2}$

3.19) Evaluate the surface integral $\iint\limits_S (x+y)\,dS$

$\mathbf{r}(u,v) = 5\cos(u)\,\mathbf{i} + 5\sin u\,\mathbf{j} + v\,\mathbf{k}$

where $0 \le u \le \dfrac{\pi}{2}, 0 \le v \le 4$.

3.19) Evaluate the surface integral $\iint\limits_{S} (x+y)\, dS$

$$\mathbf{r}(u,v) = 5\cos(u)\ \mathbf{i} + 5\sin u\ \mathbf{j} + v\ \mathbf{k}$$

where $0 \le u \le \dfrac{\pi}{2},\ 0 \le v \le 4.$

Surface Integral

$$\iint\limits_{S} (x+y)\, dS$$

S parametrized $r(u,v) = \langle 5\cos u,\ 5\sin u,\ v \rangle$
$0 \le u \le \frac{\pi}{2}$ $0 \le v \le 4$

$$ds = \| r_u \times r_v \|\, du\, dv$$

$$r_u = \langle -5\sin u,\ 5\cos u,\ 0 \rangle$$
$$r_v = \langle 0,\ 0,\ 1 \rangle$$

$$r_u \times r_v = \begin{vmatrix} i & j & k \\ -5\sin u & 5\cos u & 0 \\ 0 & 0 & 1 \end{vmatrix} = \langle 5\cos u,\ 5\sin u,\ 0 \rangle$$

$$\| r_u \times r_v \| = \sqrt{25\cos^2 u + 25\sin^2 u} = \sqrt{25(\cos^2 u + \sin^2 u)} = 5$$

$$\iint\limits_{S} (x+y)\, dS = \int_{0}^{4}\int_{0}^{\pi/2} (5\cos u + 5\sin u) \cdot 5\, du\, dv$$

$$= 25 \int_{0}^{4} (\sin u - \cos u)\Big|_{0}^{\pi/2}\, dv = 25\Big[(1-0)-(0-1)\Big]\int_{0}^{4} dv$$

$$= 25 \cdot 2 \cdot [v]_{0}^{4} = 25 \cdot 2 \cdot 4 = \boxed{200}$$

3.20) Evaluate the flux surface integral $\iint\limits_{S} \mathbf{F} \cdot \mathbf{n} \; dS$

where $\mathbf{F} = (x + y) \; \mathbf{i} + y \; \mathbf{j} + z \; \mathbf{k}$ and S is the closed surface $z = 4 - x^2 - y^2$ and $z = 0$.

3.20) Evaluate the flux surface integral $\iint\limits_{S} \mathbf{F} \cdot \mathbf{n}\ dS$

where $\mathbf{F} = (x+y)\ \mathbf{i} + y\ \mathbf{j} + z\ \mathbf{k}$ and S is the closed

surface $z = 4 - x^2 - y^2$ and $z = 0$.

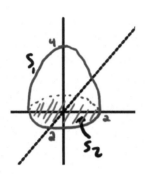

$\iint\limits_{S} F \cdot N\ ds = \iint\limits_{S_1} F \cdot N_1 dS_1 + \iint\limits_{S_2} F \cdot N_2 dS_2 = 24\pi + 0 = \boxed{24\pi}$

Flux of F over S

$\quad\quad\quad\quad\quad\quad\quad\quad\uparrow\quad\quad\quad\uparrow$

$\quad\quad\quad\quad\quad\quad\quad\quad$ paraboloid $\quad$ bottom

$S_1: z = 4 + x^2 - y^2$

$\quad$ $z = g(x,y)$ explicit

$\quad$ let $G = z - g(x,y)$

$G: z + x^2 + y^2 - 4$

$\nabla G = \langle 2x, 2y, 1 \rangle$ $\quad$ N is unit

$N_1 = \dfrac{\nabla G}{\|\nabla G\|} = \dfrac{\langle 2x, 2y, 1 \rangle}{\sqrt{4x^2 + 4y^2 + 1}}$

$ds = \sqrt{1 + (g_x)^2 + (g_y)^2}\ dA$

$dS_1 = \sqrt{1 + 4x^2 + 4y^2}\ dA$

$F = \langle x+y, y, z \rangle = \langle x+y, y, 4-x^2-y^2 \rangle$

$\left\{\begin{array}{l} z: z = 0 \\ N = \langle 0,0,-1 \rangle \text{ downward unit normal} \end{array}\right.$ $\quad F = \langle x+y, y, 0 \rangle \quad F \cdot N = 0+0+0 = \boxed{0}$

$N_1 dS_1 = \dfrac{\langle 2x, 2y, 1 \rangle}{\sqrt{4x^2+4y^2+1}} \sqrt{1 + 4x^2 + 4y^2}\ dA$

$N_1 dS_1 = \langle 2x, 2y, 1 \rangle\ dA$

$F \cdot N_1 dS_1 = (2x^2 + 2xy + 2y^2 + 4 - x^2 - y^2)\ dA$

$F \cdot N_1 dS_1 = (x^2 + y^2 + 2xy + 4)\ dA$

$\iint\limits_{S_1} F \cdot N_1\ dS_1 = \iint\limits_{R} \underbrace{(x^2+y^2}_{r^2} + 2xy + 4)\ dA$ $\quad R: x^2+y^2 = 4$ $\quad$ use double integral in polar

$\quad\quad\quad\quad\quad\quad\quad\quad\quad\quad\quad\quad\quad\quad\quad\quad x = r\cos\theta \quad\quad 0 \le \theta \le 2\pi$

$\quad\quad\quad\quad\quad\quad\quad\quad\quad\quad\quad\quad\quad\quad\quad\quad y = r\sin\theta \quad\quad 0 \le r \le 2$

$= \int_0^{2\pi} \int_0^2 (r^2 + 2r\cos\theta\ r\sin\theta + 4)\ r\ dr\ d\theta$

$= \int_0^{2\pi} \int_0^2 (r^3 + 2r^3 \cos\theta\sin\theta + 4r)\ dr\ d\theta = \int_0^{2\pi} \left(\dfrac{r^4}{4} + \dfrac{r^4}{2}\cos\theta\sin\theta + 2r^2 \right)\Big|_0^2 d\theta$

$= \int_0^{2\pi} (4 + 8\cos\theta\sin\theta + 8)\ d\theta = \int_0^{2\pi} (12 + 8\cos\theta\sin\theta)\ d\theta$ $\quad\quad\begin{array}{l} u = \sin\theta \\ du = \cos\theta\ d\theta \\ \int 8u\ du \\ \quad 4u^2 \end{array}$

$= \left[12\theta + 4(\sin\theta)^2 \right]_0^{2\pi} = \boxed{24\pi}$

3.21) Let S be the square with vertices

$(1,0,0),(0,1,0),\left(0,1,\sqrt{2}\right),$ and $\left(1,0,\sqrt{2}\right),$

and let C be the boundary of S traversed in the given order of vertices.

Let $\mathbf{W}$ be the vector field $\mathbf{W} = z\mathbf{i} + x\mathbf{j} + y\mathbf{k}$. Calculate $\int_C \mathbf{W} \cdot \mathbf{dr}$

3.21) Let S be the square with vertices

$(1,0,0), (0,1,0), (0,1,\sqrt{2})$, and $(1,0,\sqrt{2})$,

and let C be the boundary of S traversed in the given order of vertices.

Let **W** be the vector field $\mathbf{W} = z\mathbf{i} + x\mathbf{j} + y\mathbf{k}$. Calculate $\int_C \mathbf{W} \cdot \mathbf{dr}$

$\int_C W \cdot dr$

closed? yes
Ind. of Path ? No
$Q_x \neq P_y$
$R_x \neq P_z$ $Q_z \neq R_y$

Parametrize C
and find $\int_C W \, dr$

$\rightarrow$ use Stokes' Thm
$\oint_C F \cdot dr = \iint_S (\text{curl } F) \cdot n \, dS$

① Find the curl $W = \begin{vmatrix} i & j & k \\ \frac{\partial}{\partial x} & \frac{\partial}{\partial y} & \frac{\partial}{\partial z} \\ z & x & y \end{vmatrix} = \langle 1, 1, 1 \rangle$

② Find $\vec{n}$, the unit normal vector
S is a plane with no equation given
$(1,0,0)$ to $(0,1,0)$ $V = \langle -1, 1, 0 \rangle$ } cross two
$(0,1,0)$ to $(0,1,\sqrt{2})$ $W = \langle 0, 0, \sqrt{2} \rangle$ } vectors in the plane

$V \times W = \begin{vmatrix} i & j & k \\ -1 & 1 & 0 \\ 0 & 0 & \sqrt{2} \end{vmatrix} = \langle \sqrt{2}, \sqrt{2}, 0 \rangle$

$|V \times W| = \sqrt{2+2+0} = \sqrt{4} = 2$

$\vec{n} = \langle \frac{\sqrt{2}}{2}, \frac{\sqrt{2}}{2}, 0 \rangle$ OR $\frac{\sqrt{2}}{2} \langle 1, 1, 0 \rangle$

③ $\text{curl } W \cdot n = \frac{\sqrt{2}}{2} + \frac{\sqrt{2}}{2} = \sqrt{2}$

When $\text{curl } F \cdot n = \text{constant}$,
You don't need to find dS

$\iint_S (\text{curl } v \cdot n) \, dS = \sqrt{2} \cdot \boxed{\iint_S dS}$
$\sqrt{2}$

Surface area
$\sqrt{2} \times \sqrt{2}$
square
$A = 2$

$\boxed{\text{Answer} = 2\sqrt{2}}$

3.22) Find the outward flux $\iint\limits_{S} \mathbf{F} \cdot \mathbf{n}\,dS$

of the vector field $\mathbf{F} = 4xy^2\mathbf{i} + 3y\mathbf{j} + 4zx^2\mathbf{k}$ where

the surface S is the boundary of the region $1 \le x^2 + y^2 \le 4, \ 0 \le z \le 1$.

3.22) Find the outward flux $\iint\limits_S \mathbf{F}\cdot\mathbf{n}\,dS$

of the vector field $\mathbf{F} = 4xy^2\mathbf{i}+3y\mathbf{j}+4zx^2\mathbf{k}$ where

the surface S is the boundary of the region $1\leq x^2+y^2\leq 4,\ 0\leq z\leq 1$.

$$\iint\limits_S \mathbf{F}\cdot\mathbf{n}\,dS \overset{\text{Div.}}{\underset{\text{Thm.}}{=}} \iiint\limits_D \operatorname{div}\mathbf{F}\,dv$$

$$\operatorname{div}\mathbf{F} = 4y^2+3+4x^2 = \underbrace{4(x^2+y^2)}_{r^2}+3$$

use cylindrical coordinates

$0\leq z\leq 1$
$1\leq r\leq 2$
$0\leq \theta\leq 2\pi$

$$= \int_0^{2\pi}\int_1^2\int_0^1 (4r^2+3)\,r\,dz\,dr\,d\theta$$

$$= \int_0^{2\pi}\int_1^2 (4r^3+3r)\underbrace{[z]_0^1}_{1}\,dr\,d\theta = \int_0^{2\pi}\left(r^4+\frac{3r^2}{2}\right)\Big|_1^2\,d\theta$$

$$= \int_0^{2\pi}\left[\underbrace{(16+6)}_{22}-\underbrace{(1+\tfrac{3}{2})}_{-\frac{5}{2}}\right]d\theta = \frac{39}{2}\int_0^{2\pi}d\theta = \frac{39[\theta]_0^{2\pi}}{2} = \boxed{39\pi}$$

3.23) Compute the outward flux of $\nabla \times \mathbf{F}$ through the surface of the ellipsoid $2x^2 + 2y^2 + z^2 = 8$ lying above the plane $z = 0$, where $\mathbf{F} = (3x - y)\mathbf{i} + (x + 3y)\mathbf{j} + (1 + x^2 + y^2 + z^2)\mathbf{k}$.

a) 0
b) 3π
c) 12π
d) 2π
e) 8π
f) 16π

3.23) Compute the outward flux of $\nabla \times \mathbf{F}$ through the surface of the ellipsoid $2x^2 + 2y^2 + z^2 = 8$ lying above the plane $z = 0$, where $\mathbf{F} = (3x - y)\mathbf{i} + (x + 3y)\mathbf{j} + (1 + x^2 + y^2 + z^2)\mathbf{k}$.

a) 0   d) 2π
b) 3π e) 8π
c) 12π f) 16π

Compute flux of $\nabla \times F$
&
S is a surface with a boundary
curve C
$\Bigg\}$ Use Stokes' Theorem
$$\oint_C \mathbf{F} \cdot d\mathbf{r} = \iint_S (\text{curl } \mathbf{F}) \cdot \mathbf{n} \, dS$$
$\leftarrow$

① Get a good drawing

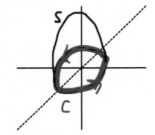

② Parametrize C
C: the intersection of
$2x^2 + 2y^2 + z^2 = 8$
and
$z = 0$
$\Rightarrow 2x^2 + 2y^2 = 8$
$x^2 + y^2 = 4$
circle of radius 2
centered at $(0,0)$

$x = 2\cos t$
$y = 2\sin t$ $0 \le t \le 2\pi$
$z = 0$

$r = \langle 2\cos t, 2\sin t, 0 \rangle$
$dr = \langle -2\sin t, 2\cos t, 0 \rangle$

③ Rewrite F on C:
$F = \langle 6\cos t - 2\sin t, 2\cos t + 6\sin t, 5 \rangle$

⑤ $\displaystyle\int_C F \cdot dr = \int_0^{2\pi} 4 \, dt$

$= \boxed{8\pi}$

④ Find $F \cdot dr$
$F \cdot dr = -12\sin t\cos t + 4\sin^2 t + 4\cos^2 t + 12\sin t\cos t = 4$

3.24) Compute the outward flux of $\mathbf{F}$ across S if

$\mathbf{F}=\left(3xy^2\right)\mathbf{i}+\left(xe^z\right)\mathbf{j}+\left(z^3\right)\mathbf{k}$ and S is the surface

of the solid bounded by the cylinder $y^2+z^2=1$

and the planes $x=-1$ and $x=2$

a) 0

b) $\dfrac{-\pi}{4}$

c) $\dfrac{11\pi}{8}$

d) 3π

e) $\dfrac{9\pi}{5}$

f) $\dfrac{9\pi}{2}$

3.24) Compute the outward flux of **F** across S if

$\mathbf{F} = (3xy^2)\mathbf{i} + (xe^z)\mathbf{j} + (z^3)\mathbf{k}$ and S is the surface

of the solid bounded by the cylinder $y^2 + z^2 = 1$

and the planes $x = -1$ and $x = 2$

a) 0   d) 3π

b) $\dfrac{-\pi}{4}$ e) $\dfrac{9\pi}{5}$

c) $\dfrac{11\pi}{8}$ f) $\dfrac{9\pi}{2}$

Compute outward flux of F
&
S is the boundary of a solid 3D Region

$\left.\rule{0pt}{40pt}\right\}$ Use the Divergence Theorem

$$\iint_S (\mathbf{F} \cdot \mathbf{n})\, dS = \iiint_D div\, \mathbf{F}\, dV = \int_0^{2\pi}\int_0^1\int_{-1}^2 3r^2\, dx\, r\, dr\, d\theta$$

① Find $div\, F = P_x + Q_y + R_z$ when $F = \langle P, Q, R\rangle$

$\quad div\, F = 3y^2 + 0 + 3z^2 = 3(y^2 + z^2)$

② Draw the region find the bounds

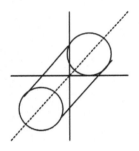

Cylindrical but with X acting like Z $y^2 + z^2 = r^2$

$dv = dx\, r\, dr\, d\theta$

$-1 \leq x \leq 2$
$0 \leq r \leq 1$
$0 \leq \theta \leq 2\pi$

$$= \int_0^{2\pi}\int_0^1 3r^3 [x]_{-1}^2\, dr\, d\theta$$

$$2 - (-1) = 3$$

$$= 9\int_0^{2\pi}\int_0^1 r^3\, dr\, d\theta$$

$$= 9\left[\frac{r^4}{4}\right]_0^1 \cdot \int_0^{2\pi} d\theta$$

$$= \frac{9}{4} \cdot 2\pi = \boxed{\dfrac{9\pi}{2}}$$

3.25) Find the outward flux $\iint\limits_S \mathbf{F}\cdot\mathbf{n}\,dS$

of the vector field $\mathbf{F} = 2xz\,\mathbf{i} + 5y^2\,\mathbf{j} - z^2\mathbf{k}$ where
the surface S is the boundary of the solid pictured below.

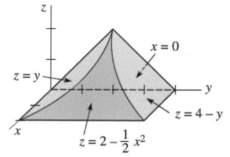

a) 232   b) 128   c) 112   d) 96   e) 72   f) 64   g) 40   h) 36

3.25) Find the outward flux $\iint\limits_S \mathbf{F} \cdot \mathbf{n}\, dS$

of the vector field $\mathbf{F} = 2xz\,\mathbf{i} + 5y^2\,\mathbf{j} - z^2\mathbf{k}$ where
the surface S is the boundary of the solid pictured below.

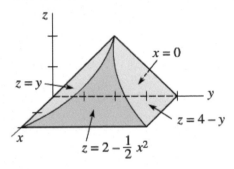

a) 232   b) 128   c) 112   d) 96   e) 72   f) 64   g) 40   h) 36

Compute outward flux of F
&
S is the boundary of a solid 3D Region

Use the Divergence Theorem
$$\iint_S (\mathbf{F} \cdot \mathbf{n})\, dS = \iiint_D \operatorname{div} \mathbf{F}\, dV = \int_0^2 \int_0^{2-\frac{1}{2}x^2} \int_z^{4-z} 10y\, dy\, dz\, dx$$

① Find $\operatorname{div} F = P_x + Q_y + R_z$ when $F = \langle P, Q, R\rangle$
$$\operatorname{div} F = 2z + 10y - 2z = 10y$$

$$= \int_0^2 \int_0^{2-\frac{1}{2}x^2} 5y^2 \Big]_z^{4-z} dz\, dx$$

② Figure out the bounds
z first? No   x first?

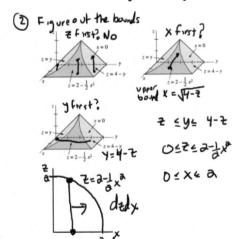

upper bound $x = \sqrt{4-z}$

y first?

$z \le y \le 4-z$

$y = 4-z$

$0 \le z \le 2 - \frac{1}{2}x^2$

$z = 2 - \frac{1}{2}x^2$

$0 \le x \le 2$

$dz\, dx$

$$= 5\int_0^2 \int_0^{2-\frac{1}{2}x^2} \underbrace{(4-z)^2 - z^2}_{16-8z+z^2-z^2}\, dz\, dx$$

$$= 5\int_0^2 \int_0^{2-\frac{1}{2}x^2} (16-8z)\, dz\, dx$$

$$= 5\int_0^2 \Big[16z - 4z^2\Big]_0^{2-\frac{1}{2}x^2} dx$$

$$= 20\int_0^2 (4z - z^2)\Big|_0^{2-\frac{1}{2}x^2} dx$$

$$\underset{4(2-\frac{1}{2}x^2) - (2-\frac{1}{2}x)^2}{}$$

$$= 20\int_0^2 \underbrace{8 - 2x^2 - [4 - 2x^2 + \frac{x^4}{4}]}_{4 - \frac{x^4}{4}}\, dx$$

$$= \int_0^2 (80 - 5x^4)\, dx = 80x - x^5\Big|_0^2$$

$$= 160 - 32 = \boxed{128}$$

3.26) Determine the value of the line integral $\int_C \mathbf{F} \cdot d\mathbf{r}$ where

$\mathbf{F} = (x + 2z) \; \mathbf{i} + (3x + y) \; \mathbf{j} + (2y - z) \mathbf{k}$

and C is the curve of the intersection of the plane $x + 2y + z = 4$

with the coordinate planes oriented counterclockwise as viewed from above.

3.26) Determine the value of the line integral $\int_C \mathbf{F} \cdot d\mathbf{r}$ where

$$\mathbf{F} = (x+2z)\ \mathbf{i} + (3x+y)\ \mathbf{j} + (2y-z)\mathbf{k}$$

and C is the curve of the intersection of the plane $x + 2y + z = 4$

with the coordinate planes oriented counterclockwise as viewed from above.

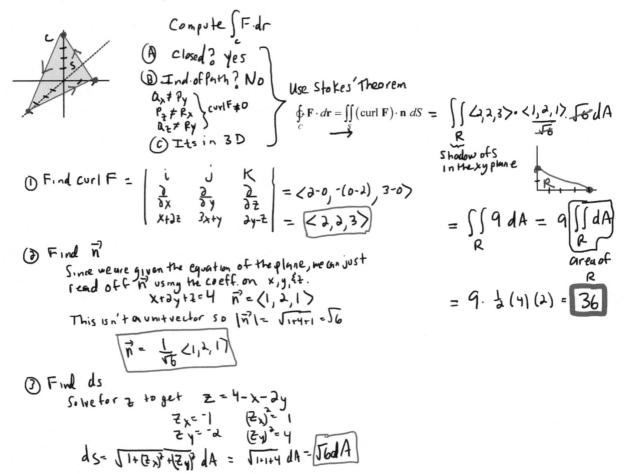

Compute $\int_c F \cdot dr$

Ⓐ closed? yes

Ⓑ Ind. of Path? No

$a_y \neq P_y$
$P_z \neq R_x$ } curl F ≠ 0
$a_z \neq R_y$

Ⓒ Its in 3D

Use Stokes' Theorem

$\oint_c F \cdot dr = \iint_S (\text{curl } F) \cdot n\, dS = \iint_R \langle 2,2,3 \rangle \cdot \langle 1,2,1 \rangle \dfrac{\sqrt{6}}{\sqrt{6}} dA$

R
↑
Shadow of S
in the xy plane

Ⓞ Find curl F = $\begin{vmatrix} i & j & k \\ \frac{\partial}{\partial x} & \frac{\partial}{\partial y} & \frac{\partial}{\partial z} \\ x+2z & 3x+y & 2y-z \end{vmatrix}$ $= \langle 2-0, -(0-2), 3-0 \rangle$
$= \boxed{\langle 2,2,3 \rangle}$

$= \iint_R 9\, dA = 9 \iint_R dA$

area of
R

② Find $\vec{n}$

Since we are given the equation of the plane, we can just read off $\vec{n}$ using the coeff. on x, y, & z.

$x + 2y + z = 4$ $\vec{n} = \langle 1, 2, 1 \rangle$

This isn't a unit vector so $|\vec{n}| = \sqrt{1+4+1} = \sqrt{6}$

$= 9 \cdot \frac{1}{2} (4)(2) = \boxed{36}$

$\boxed{\vec{n} = \frac{1}{\sqrt{6}} \langle 1,2,1 \rangle}$

③ Find ds

Solve for z to get $z = 4 - x - 2y$

$z_x = -1$ $(z_x)^2 = 1$
$z_y = -2$ $(z_y)^2 = 4$

$ds = \sqrt{1 + (z_x)^2 + (z_y)^2}\, dA = \sqrt{1+1+4}\, dA = \boxed{\sqrt{6}\, dA}$

3.27) Compute the outward flux of curl $\mathbf{F}$ across S if

$$\mathbf{F} = (4y\cos z)\mathbf{i} + (e^x \sin z)\mathbf{j} + (xe^y)\mathbf{k}$$

and S is the upper hemisphere of $x^2 + y^2 + z^2 = 25$, oriented downward.

a) 25π d) 100π

b) 50π e) 120π

c) 80π f) None of these

3.27) Compute the outward flux of curl $\mathbf{F}$ across S if

$$\mathbf{F} = (4y\cos z)\mathbf{i} + (e^x \sin z)\mathbf{j} + (xe^y)\mathbf{k}$$

and S is the upper hemisphere of $x^2 + y^2 + z^2 = 25$, oriented downward.

a) 25π d) 100π
b) 50π e) 120π
c) 80π f) None of these

$c: \; x = 5\cos t \qquad r = \langle 5\cos t, 5\sin t, 0\rangle$
$\qquad y = 5\sin t$
$\qquad z = 0 \qquad dr = \langle -5\sin t, 5\cos t, 0\rangle$
$\qquad t \; st. \; at \; 2\pi$
$\qquad and \; ends \; at \; 0 \quad F_{on \; c} = \langle 4\cdot 5\sin t \cos 0, e^{5\cos t} \sin 0, 5\cos t \, e^0\rangle$

$$F_{on \; c} = \langle 20\sin t, 0, 5\cos t\rangle$$

$$F \cdot dr = -100\sin^2 t + 0 + 0$$

$$\iint_S \text{curl} F \cdot n \, ds \overset{\text{Stokes'}}{\underset{\text{Thm}}{=}} \oint_c F \, dr = \int_{2\pi}^{0} -100\sin^2 t \, dt = 100\int_0^{2\pi} \tfrac{1}{2}(1 - \cos 2t) \, dt$$

$$= 50\left[t - \tfrac{1}{2}\sin(2t)\right]_0^{2\pi} = \boxed{100\pi}$$

3.28) Determine the value of the line integral $\int_C \mathbf{F} \cdot d\mathbf{r}$ where

$$\mathbf{F} = \left(xy^2\right)\mathbf{i} + \left(\frac{x}{z}\right)\mathbf{j} + \left(2x+y\right)\mathbf{k}$$

and C is the rectangular path with vertices $(0,0,2),(0,3,2),(4,3,2),(4,0,2)$, and $(0,0,2)$ in that order.

3.28) Determine the value of the line integral $\int_C \mathbf{F} \cdot d\mathbf{r}$ where

$$\mathbf{F} = \left(xy^2\right)\mathbf{i} + \left(\frac{x}{z}\right)\mathbf{j} + \left(2x + y\right)\mathbf{k}$$

and C is the rectangular path with vertices $(0,0,2),(0,3,2),(4,3,2),(4,0,2)$, and $(0,0,2)$ in that order.

$$\int_C \mathbf{F} \cdot d\mathbf{r} \overset{\text{Stokes' Thm}}{=} \int\int_S \text{curl} \mathbf{F} \cdot \mathbf{n} \, ds$$

$\int_C$ Clockwise

$S: z=2 \quad \vec{n}=\langle 0,0,-1\rangle$ Since Clockwise

$z_x = 0 \quad z_y = 0$

$ds = \sqrt{1+(z_x)^2+(z_y)^2} \, dA \Rightarrow ds = dA$

$$\text{curl } \mathbf{F} = \begin{vmatrix} \mathbf{i} & \mathbf{j} & \mathbf{K} \\ \frac{\partial}{\partial x} & \frac{\partial}{\partial y} & \frac{\partial}{\partial z} \\ xy^2 & \frac{x}{z} & 2x+y \end{vmatrix} = \langle 1-\frac{x}{z^2}, -(2-0), \frac{1}{z}-2xy\rangle$$

$\text{curl} \mathbf{F} \cdot \mathbf{n} = 0 + 0 - (\frac{1}{z} - 2xy)$

$z=2$

$\text{curl} \mathbf{F} \cdot \mathbf{n} = 2xy - \frac{1}{2}$

$$= \int\int_R (2xy - \tfrac{1}{2}) \, dA = \int_0^4 \int_0^3 (2xy - \tfrac{1}{2}) \, dy \, dx = \int_0^4 \left(xy^2 - \tfrac{1}{2}y\right)_0^3 \, dx$$

$$= \int_0^4 (9x - \tfrac{3}{2}) \, dx = \left[\frac{9x^2}{2} - \frac{3}{2}x\right]_0^4 = (9 \cdot 8 - 6) - 0 = \boxed{66}$$

$72 - 6$

3.29) Find the outward flux $\iint\limits_{S} \mathbf{F} \cdot \mathbf{n} dS$ of the vector field

$\mathbf{F} = x\sqrt{x^2 + y^2 + z^2}\ \mathbf{i} + y\sqrt{x^2 + y^2 + z^2}\ \mathbf{j} + z\sqrt{x^2 + y^2 + z^2}\ \mathbf{k}$

where the surface S consists of the hemisphere

$z = \sqrt{1 - x^2 - y^2}$ and the unit disk in the xy-plane $x^2 + y^2 \leq 1$.

a) 0

d) 2π

b) 3π

e) 8π

c) 12π

f) 16π

3.29) Find the outward flux $\iint_S \mathbf{F} \cdot \mathbf{n}\, dS$ of the vector field

$$\mathbf{F} = x\sqrt{x^2 + y^2 + z^2}\ \mathbf{i} + y\sqrt{x^2 + y^2 + z^2}\ \mathbf{j} + z\sqrt{x^2 + y^2 + z^2}\ \mathbf{k}$$

where the surface S consists of the hemisphere

$z = \sqrt{1 - x^2 - y^2}$ and the unit disk in the xy – plane $x^2 + y^2 \leq 1$.

a) 0   d) 2π

b) 3π e) 8π

c) 12π f) 16π

Closed
Bounded
Surface
S

$$\iint_S F \cdot n\, ds \xrightarrow[\text{Thm.}]{\text{Div.}} \iiint_D \mathrm{div}F\, dV$$

$$\mathrm{div}\,F = \underbrace{\sqrt{x^2+y^2+z^2}} + x \cdot \frac{1}{2\sqrt{x^2+y^2+z^2}} \cdot 2x + \sqrt{x^2+y^2+z^2} + y \cdot \frac{1}{2\sqrt{x^2+y^2+z^2}} \cdot 2y + \underbrace{\sqrt{x^2+y^2+z^2} + z \cdot \frac{1}{2\sqrt{x^2+y^2+z^2}} \cdot 2z}$$

$$\mathrm{div}F = 3\sqrt{x^2+y^2+z^2} + \frac{x^2+y^2+z^2}{\sqrt{x^2+y^2+z^2}} = 4\sqrt{x^2+y^2+z^2} = 4\rho$$

use spherical $0 \leq \rho \leq 1$
$0 \leq \phi \leq \pi/2$
$0 \leq \theta \leq 2\pi$

$$= \int_0^{2\pi} \int_0^{\pi/2} \int_0^1 4\rho \cdot \rho^2 \sin\phi\, d\rho\, d\phi\, d\theta = \int_0^{2\pi} \int_0^{\pi/2} \left[\rho^4\right]_0^1 \sin\phi\ d\phi\, d\theta$$

$$= \int_0^{2\pi} \int_0^{\pi/2} \sin\phi\, d\phi\ d\theta = \int_0^{2\pi} \left[-\cos\phi\right]_0^{\pi/2}\, d\theta = \int_0^{2\pi} d\theta = \boxed{2\pi}$$

$(0--1)$

3.30) Determine the value of the line integral $\int_C \mathbf{F}\cdot d\mathbf{r}$ where

$$\mathbf{F} = \left(\sin x - \frac{y^3}{3}\right)\mathbf{i} + \left(\cos y + \frac{x^3}{3}\right)\mathbf{j} + (xyz)\mathbf{k}$$

and C is the intersection between the cone $z^2 = x^2 + y^2$ and the plane $z = 1$ traversed counter-clockwise.

A) $\dfrac{\pi}{3}$ B) $\dfrac{\pi}{6}$ C) $\dfrac{\pi}{4}$ D) $\dfrac{\pi}{2}$

E) $\dfrac{4\pi}{3}$ F) $\dfrac{2\pi}{3}$ G) π H) None of these

3.30) Determine the value of the line integral $\int_C \mathbf{F} \cdot d\mathbf{r}$ where

$$\mathbf{F} = \left(\sin x - \frac{y^3}{3} \right) \mathbf{i} + \left(\cos y + \frac{x^3}{3} \right) \mathbf{j} + (xyz) \mathbf{k}$$

and C is the intersection between the cone $z^2 = x^2 + y^2$

and the plane $z = 1$ traversed counter-clockwise.

A) $\dfrac{\pi}{3}$ B) $\dfrac{\pi}{6}$ C) $\dfrac{\pi}{4}$ D) $\dfrac{\pi}{2}$

E) $\dfrac{4\pi}{3}$ F) $\dfrac{2\pi}{3}$ G) π H) None of these

$$\int_C \mathbf{F} \cdot d\mathbf{r} \overset{\text{Stokes'}}{\underset{\text{Thm}}{=\!=\!=}} \iint_S \mathbf{F} \cdot \mathbf{n} \, ds$$

$n = \langle 0, 0, 1 \rangle$

$z = 1$

$z_x = 0 \quad z_y = 0$

$ds = \sqrt{1 + (z_x)^2 + (z_y)^2} \, dA \Rightarrow ds = dA$

$$\text{curl}\,\mathbf{F} = \begin{vmatrix} \mathbf{i} & \mathbf{j} & \mathbf{k} \\ \frac{\partial}{\partial x} & \frac{\partial}{\partial y} & \frac{\partial}{\partial z} \\ \sin x - \frac{y^3}{3} & \cos y + \frac{x^3}{3} & xyz \end{vmatrix} = \langle xz - 0, -(yz - 0), x^2 - -y^2 \rangle$$

$= \langle xz, -yz, x^2 + y^2 \rangle$

$\text{curl}\,\mathbf{F} \cdot \mathbf{n} \, ds = (0 + 0 + x^2 + y^2) \, dA = (x^2 + y^2) dA$

use polar r^2

$$\int_0^{2\pi} \int_0^1 r^2 \cdot r \, dr \, d\theta = \int_0^{2\pi} \left[\frac{r^4}{4} \right]_0^1 d\theta = \frac{1}{4} [\theta]_0^{2\pi} = \boxed{\frac{\pi}{2}}$$

$(2\pi - 0)$

3.31) Use the Divergence Theorem to evaluate the surface integral $\iint\limits_{S} \mathbf{F} \cdot \mathbf{n} \; dS$ where $\mathbf{F} = \langle x + \cos y, \ln z, x^2 \rangle$ and S is the surface of the

hemisphere $x^2 + y^2 + z^2 = 1$ with $z > 0$ and $\mathbf{n}$ is the outward normal to S.

A) $\dfrac{7\pi}{6}$ B) $\dfrac{11\pi}{12}$ C) $\dfrac{5\pi}{4}$ D) $\dfrac{5\pi}{12}$

E) $\dfrac{7\pi}{3}$ F) $\dfrac{5\pi}{2}$ G) $\sqrt{2}\pi$ H) None of these

3.31) Use the Divergence Theorem to evaluate the surface integral

$\iint\limits_{S} \mathbf{F} \cdot \mathbf{n} \, dS$ where $\mathbf{F} = \langle x + \cos y, \ln z, x^2 \rangle$ and S is the surface of the

hemisphere $x^2 + y^2 + z^2 = 1$ with $z > 0$ and $\mathbf{n}$ is the outward normal to S.

A) $\dfrac{7\pi}{6}$ B) $\dfrac{11\pi}{12}$ C) $\dfrac{5\pi}{4}$ D) $\dfrac{5\pi}{12}$

E) $\dfrac{7\pi}{3}$ F) $\dfrac{5\pi}{2}$ G) $\sqrt{2}\pi$ H) None of these

$$\iint_{S} F \cdot n \, ds + \iint_{S_1} F \cdot n_1 \, ds_1 \overset{\text{Div.}}{\underset{\text{Thm.}}{=\!=}} \iiint_{D} \text{div} \, F \, dV$$

$$\iint_{S} F \cdot n \, ds = \iiint_{D} dv - \iint_{S_1} F \cdot n_1 \, ds_1,$$

$$\underbrace{}_{\text{Volume}}$$

$\text{div } F = 1 + 0 + 0 = 1$

$\frac{1}{2}(\frac{4}{3}\pi r^3)$

$\frac{2}{3}\pi$

$n_1 = \langle 0, 0, -1 \rangle$

$F \cdot n_1 = 0 + 0 + -x^2$

$S: z = 0$

$z_x = 0$ $ds = \sqrt{1 + (z_x)^2 + (z_y)^2} \, dA$

$z_y = 0$

$ds = dA$

$= \dfrac{2\pi}{3} \qquad\qquad -\iint_{R} -x^2 \, dA$

polar

$= \dfrac{2\pi}{3} + \iint_{0}^{2\pi}\int_{0}^{1} r^2 \cos^2\theta \, r \, dr \, d\theta = \dfrac{2\pi}{3} + \int_{0}^{2\pi} \left[\frac{r^4}{4}\right]_{0}^{1} \cos^2\theta \, d\theta$

$= \dfrac{2\pi}{3} + \frac{1}{4} \cdot \frac{1}{2} \int_{0}^{2\pi} (1 + \cos 2\theta) \, d\theta = \dfrac{2\pi}{3} + \frac{1}{8} \left[\theta + \frac{1}{2}\sin(2\theta)\right]\Big|_{0}^{2\pi}$

$= \dfrac{2\pi}{3} + \frac{1}{8}(2\pi) = \dfrac{2\pi}{3} + \frac{\pi}{4} = \dfrac{8\pi + 3\pi}{12} = \boxed{\dfrac{11\pi}{12}}$

3.32) Consider the vector field $\mathbf{W} = x^3 y^2 \mathbf{i} - x^2 y^3 \mathbf{j} + (1+z)\mathbf{k}$.

Find the outward flux of $\mathbf{W}$ through the portion S of the paraboloid

$z = 4 - x^2 - y^2$ which lies above the xy-plane

3.32) Consider the vector field $\mathbf{W} = x^3y^2\mathbf{i} - x^2y^3\mathbf{j} + (1+z)\mathbf{k}$.

Find the outward flux of $\mathbf{W}$ through the portion S of the paraboloid

$z = 4 - x^2 - y^2$ which lies above the xy-plane

Find $\iint_{S_1}(w \cdot N)\,dS \Rightarrow$ Add the surface integral over S_2 to close the Surface.

• This allows you to use the divergence theorem.

$$\iint_{S_1}(w \cdot N_1)\,dS_1 + \iint_{S_2}(w \cdot N_2)\,dS_2 \overset{\text{Div.Thm.}}{=} \iiint_D \operatorname{div}W\,dV$$

N_2 points downward

Right hand rule $\Rightarrow S_2$ has clockwise orientation

$$\iint_{S_1}(w \cdot n)\,dS_1 = \iiint_D \operatorname{div}W\,dV - \iint_{\substack{S_2 \\ \text{clockwise}}}(w \cdot N_2)\,dS_2$$

$$\iint_{S_1}(w \cdot N_1)\,dS_1 = \iiint_D (3x^2y^2 - 3x^2y^2 + 1)\,dv + \iint_{\substack{S_2 \\ \text{counter} \\ \text{clockwise}}}(w \cdot \vec{n})\,dS_2$$

$\vec{n} = \langle 0,0,1\rangle$
$\Rightarrow$ upward normal

$w \cdot \vec{n} = 0+0+1\cdot z$
but $z = 0$
$w \cdot \vec{n} = 1$

$$\iint_{S_1}(w \cdot N_1)\,dS_1 = \int_0^{2\pi}\int_0^2\int_0^{4-r^2} dz\, r\, dr\, d\theta + \iint_{S_2} 1\, dS_2$$

Surface Area Planar Surface

$A = \pi r^2$
$A = 4\pi$

$$\iint_{S_1}(w \cdot N_1)\,dS_1 = \int_0^{2\pi}\int_0^2 (4-r^2)\,r\,dr\,d\theta + 4\pi$$

$4r - r^3$

$\left(2r^2 - \frac{r^4}{4}\right)_0^2 \cdot \int_0^{2\pi} d\theta$

$(8-4) \cdot 2\pi$

$$\iint_{S_1}(w \cdot N_1)\,dS_1 = 8\pi + 4\pi = \boxed{12\pi}$$

3.33) Compute $\iint\limits_S \text{curl}\mathbf{F} \cdot \mathbf{n} \, dS$

where $\mathbf{F} = xy\mathbf{i} + (zy - 2y)\mathbf{j} + y^2 z\mathbf{k}$.

and S is the cone $z^2 = x^2 + y^2$ with $0 \le z \le 2$

and $\mathbf{n}$ the outward (downward pointing) normal.

3.33) Compute $\iint\limits_S \text{curl}\mathbf{F} \cdot \mathbf{n} \, dS$

where $\mathbf{F} = xy\mathbf{i} + (zy - 2y)\mathbf{j} + y^2 z\mathbf{k}$.

and S is the cone $z^2 = x^2 + y^2$ with $0 \le z \le 2$

and $\mathbf{n}$ the outward (downward pointing) normal.

$$\iint\limits_S \text{curl} \mathbf{F} \cdot \mathbf{n} \, dS \underset{\text{Thm.}}{\overset{\text{Stokes'}}{=}} \oint_C \mathbf{F} \cdot d\mathbf{r}$$

$$= \int_{2\pi} -8 \sin t \cos t \, dt$$

$$= 8 \int_0^{2\pi} \sin t \cos t \, dt$$

$$= 4 \int_0^{2\pi} 2\sin t \cos t \, dt$$

$$= 4 \int_0^{2\pi} \sin(2t) \, dt$$

$$= -2 \cos(2t) \Big|_0^{2\pi}$$

$$= -2 (1 - 1)$$

$$= \boxed{0}$$

clockwise
since $\vec{n}$ points
downward

S

$c:\ x = 2\cos t$
$\quad y = 2\sin t$
$\quad z = 2$
t st. at 2π ends at 0

$r = \langle 2\cos t, 2\sin t, 2 \rangle$
$dr = \langle -2\sin t, 2\cos t, 0 \rangle$
$\mathbf{F}$ on $c = \langle 4\cos t, 4\sin t - 8\sin t, 8\sin^2 t \rangle$
$\mathbf{F} \cdot d\mathbf{r} = -8\sin t \cos t$